Life-Cycle Assessment of Semiconductors

T0135124

Sarah B. Boyd

Life-Cycle Assessment of Semiconductors

Foreword by Arpad Horvath

 Springer

Sarah B. Boyd
PE International
344 Boylston Street, 3rd Floor,
Boston, MA 02116, USA
S.Boyd@pe-international.com

ISBN 978-1-4899-9223-9   ISBN 978-1-4419-9988-7 (eBook)
DOI 10.1007/978-1-4419-9988-7
Springer New York Dordrecht Heidelberg London

Printed on acid-free paper

Springer is part of Springer Science+Business Media (www.springer.com)

This work is dedicated to Jean G. Boyd

Foreword

In the 2010 Oscar-nominated movie The Social Network, much is made of the importance of brilliant ideas, computer algorithms, business models, copyrights, and other legal issues surrounding the creation and launch of one of the most popular internet companies. What is not talked about is that without the microchips and other semiconductor devices filling mushrooming data centers and demanding the electricity output of entire power plants, nothing of the social network depicted in the movie and growing larger each day would exist, and the internet economy as we know it, worth hundreds of billions of dollars would not operate. We would not have had the "internet revolution" of the last decade and a half, and we would be devoid of the delicious speculations about the unfathomable growth of the internet economy in the times to come. Oh, and this movie would not have been made.

Sarah Boyd's book fills several gaps. It provides the most comprehensive environmental assessment of semiconductor manufacturing to date. She summarizes the existing literature critically and selectively, and through her own research provides data, approaches, models, and results that complement prior work or offer original results. Both approaches are, of course, admirable, but this book is not primarily a synthesis, a meta study of prior work, annotated and improved. Sarah Boyd did not have the luxury of standing on the shoulder of giants. Rather this is dominantly a compendium of Dr. Boyd's research results that she arrived at by using lots of original ideas, insights, approaches, and data. This book provides results for products for which no results or merely approximations were available heretofore. Many data points were collected directly from company processes, and many data are of higher transparency as well as quality than seen ever before.

From a disciplinary viewpoint, the research embodied in this book links electrical, chemical, production, and environmental engineering. There is not an abundance of peer-reviewed articles in this sphere, and the number of books is even fewer.

Noteworthy is the gap that this work fills in the life-cycle assessment (LCA) methodology for semiconductor manufacturing. Due to access to data and process engineering knowledge, the book's focus is on consumption of resources on individual process level rather than on the level of an entire manufacturing facility, hence this bottom-up approach yields more focused and robust results than the previous literature that is top-down, manufacturing site down to individual processes, with unavoidable allocation procedures and uncertainties.

Very rare is an LCA book that is able to tackle so many products with such detail and over several technological generations, while not losing sight of the economic perspective and geographic specificity of the semiconductor industry, and focusing not just on manufacturing and transportation, but also on the use phase of a product. As well, this may be the first LCA book written for a product of such small size but sold at such large scale worldwide like a typical semiconductor device.

Dear Reader, a practitioner, or a scholar, happy reading of this book we can all learn much from.

Berkeley, CA, USA   Arpad Horvath

Preface

The purpose of this book is to quantify and describe the environmental impacts of semiconductor manufacturing for life-cycle assessment (LCA) practitioners and students, in order to enable more representative LCA of electronic products. This book describes the available analytical techniques for LCA of semiconductors, as well as a set of life cycle inventory (LCI) data for several semiconductor device types over many technological generations. Using the methods and data presented in the book, LCA practitioners can adapt these LCI for use in LCA of many types of electronic products.

The introductory chapter of this book provides an overview of the types of environmental impacts which occur in production of semiconductors, and relates a more detailed picture of the objectives of this book. A history of the published work on the topic of semiconductor LCA is outlined, to provide context to the LCI and LCA studies presented in the later chapters. Chapter 2 aims to provide the reader with a toolbox of LCA approaches, by providing an overview of methods applied in semiconductor LCA, including approaches used in policy as well as in academic research. The chapter includes a step-by-step guide to the application of generic semiconductor LCI in LCA of electronics, which may be useful to the reader in adapting LCI presented in Chaps. 4–7. Chapter 3 presents an economic perspective on the semiconductor industry with the purpose of providing an explanation of the major semiconductor product types as well as describing environmentally relevant trends in the geographic concentration of semiconductor manufacturing. Chapter 4 is a life-cycle energy analysis of CMOS logic over several technology generations, with a focus on primary energy consumption and GWP, as well as the sensitivity of impacts to production metrics and product performance. Chapter 5 presents an LCA for CMOS logic, including all environmental impact categories. LCA of flash memory is presented in Chap. 6. The topic of functional unit choice in semiconductor LCA and its impact on LCA results is explored in Chap. 7, in which multi-generational LCA of DRAM is presented. The final chapter provides an outlook for future research in the area of semiconductor LCA, and LCA of electronics and information technology.

Boston, MA, USA

Sarah B. Boyd

Acknowledgments

The author acknowledges the work of Sebastien Raoux, Mehran Moalem, Rafika Smati, Mark Denome and Nikhil Krishnan, whose previous work on emissions measurement and semiconductor LCA laid the foundation for this research. This work would also not have been possible without the advice and support of my advisors David Dornfeld and Arpad Horvath, as well as the review and perspectives of Al Pisano, John Harland and Andreas Neuber.

This research was supported by the Robert Noyce Fellowship, Applied Materials, the EPA STAR Grant on Comprehensive Tools to Assess Environmental Impacts of and Improve the Design of Semiconductor Equipment and Processes, and the University of California Toxic Substances Research and Teaching Program. In its early stages, guidance was also provided to this research via the National Science Foundation/Semiconductor Research Corporation Environmentally Benign Semiconductor Manufacturing Engineering Research Center.

Special thanks go to Rudi Schuech.

Contents

1   **Introduction** 1
    1.1   Semiconductor Life-Cycle Environmental Impacts 2
    1.2   Challenges in LCA of Semiconductors 4
    1.3   Timeline of Semiconductor LCA in Practice 4
    1.4   LCA of Information Technology 9
    1.5   Overview of this Book 10

2   **Semiconductor LCI Methods** 13
    2.1   LCI of Wafer Fabrication 13
          2.1.1   Mass and Energy Flow Modeling 14
          2.1.2   Government and Policy Methods Applicable
                  to Semiconductor LCI 17
          2.1.3   Adaptation of Existing Semiconductor LCI Data 20
          2.1.4   Use of Economic Input-Output LCA for Wafer
                  Fabrication 21
    2.2   LCI of Chemical Production 21
    2.3   LCI of Infrastructure 23
    2.4   Modeling Energy Consumption in the Use Phase 23
          2.4.1   Specifying the Use Phase Scenario: Location,
                  Time in Operation and Utilization Rate 24
          2.4.2   System-Level Factors 25
          2.4.3   Software 26
    2.5   End of Life 26

3   **Semiconductor Manufacturing Trends
    in Product Type and Geography** 27
    3.1   Introduction 27
    3.2   The Semiconductor Industry: Size, Growth and Trends 28
          3.2.1   Technology Scaling 28
          3.2.2   Dominant Circuit Elements: The Transition
                  from Bipolar to CMOS Transistors 29

	3.2.3 Products	31
	3.2.4 Geographic Concentration of Production	32
3.3	Conclusions	35

4 Life-Cycle Energy and Global Warming Emissions of CMOS Logic 37

4.1	Introduction	37
4.2	Methodology	38
	4.2.1 Inventory Model Structure	38
	4.2.2 "Upstream" Materials	41
	4.2.3 Infrastructure and Equipment	46
	4.2.4 Electricity	46
	4.2.5 Semiconductor Manufacturing	48
	4.2.6 Transportation	55
	4.2.7 Use Phase	55
4.3	Results and Discussion	57
	4.3.1 Uncertainty Assessment	62
	4.3.2 Sensitivity Analysis	64
	4.3.3 Data Quality	67
4.4	Discussion	68

5 Life-Cycle Assessment of CMOS Logic 69

5.1	Introduction	69
5.2	Methods	71
	5.2.1 Manufacturing Process Power and Emissions	72
	5.2.2 Facility Utility Systems and Facility Energy Efficiency	73
	5.2.3 Abatement	73
	5.2.4 Environmental Impact Metrics	75
	5.2.5 Electricity Generation Emissions Factors	75
	5.2.6 Use Phase Power	76
5.3	Results and Discussion	76
	5.3.1 Global Warming Potential	76
	5.3.2 Water Use	77
	5.3.3 Photochemical Oxidant Formation, Acidification and Eutrophication Impacts via Air Emissions	78
	5.3.4 Human Health Impacts	80
5.4	Uncertainty	83
	5.4.1 Sensitivity Analysis	83
	5.4.2 Unaccounted Impacts	84
5.5	Discussion and Conclusions	85

6 Life-Cycle Assessment of Flash Memory 87

6.1	Introduction	87
6.2	Methodology	88
6.3	Results	89

6.4   Discussion: Solid State Drives vs. Hard Disk Drives   92
        6.4.1   Uncertainty...   93
        6.4.2   Sensitivity Analysis   94
6.5   Conclusions   95

7 Life-Cycle Assessment of Dynamic Random Access Memory   97
7.1   Introduction   97
7.2   Methodology   98
7.3   Results...   99
        7.3.1   Resource Consumption.......................................   99
        7.3.2   Environmental Impacts.......................................   99
        7.3.3   Human Health Impacts   103
7.4   Discussion...   104
        7.4.1   Comparison with Existing Work.............................   104
7.5   Conclusions   107

8 Semiconductor LCA: The Road Ahead   109
8.1   Future Work   110
        8.1.1   Life-Cycle Inventory and Impact Factors
                 for Semiconductor Chemicals...............................   110
        8.1.2   LCA of Networked Thin Client Versus Local
                 Desktop Computing   111
        8.1.3   Comparative LCA of IT Versus Traditional
                 Products and Services   111
        8.1.4   Further Study of the "Rebound Effect"
                 of Computational Power and Memory Capacity   112

A CMOS Logic: Manufacturing Inventory and Impact Data............   113
A.1   CMOS Logic Manufacturing Inventory Data.......................   113
A.2   CMOS Logic Manufacturing Emissions Data.......................   125
A.3   CMOS Logic Impact Results..   142

B Flash Memory Life Cycle Inventory and Impact Results..............   151
B.1   Flash Memory: Manufacturing Inventory Data.....................   151
B.2   Flash Memory: Manufacturing Emissions Data....................   159
B.3   Flash LCA: Tabulated Results with Uncertainty   169

C DRAM Life Cycle Inventory and Impact Results   179
C.1   DRAM Manufacturing Inventory Data   179
C.2   DRAM Manufacturing Emissions Data   188
C.3   DRAM LCA: Tabulated Results with Uncertainty   201

References..   213

Index..   221

List of Figures

Fig. 1.1 Development timeline for an advanced semiconductor product 4
Fig. 1.2 Compilation of individual process models into process flows 11

Fig. 2.1 Life-cycle stages of an integrated circuit............................ 14
Fig. 2.2 Quadrupole mass spectrometry....................................... 15
Fig. 2.3 Fourier transform infrared spectroscopy 16

Fig. 3.1 Global annual revenues from 1952 to 2007......................... 29
Fig. 3.2 Worldwide MOS production capacity by technology node 30
Fig. 3.3 2007 Worldwide revenues by product type 32
Fig. 3.4 2007 Worldwide revenues by geographic region 34

Fig. 4.1 Life-cycle stages 38
Fig. 4.2 Overview of mass and energy flows considered in the
         fab model 49
Fig. 4.3 Energy use per die, per wafer and per million transistors
         by technology node 58
Fig. 4.4 Energy use per 300 mm wafer equivalent, by life-cycle
         stage, over eight technology nodes 58
Fig. 4.5 Energy use per die, by life-cycle stage, over eight
         technology nodes... 59
Fig. 4.6 GWP per die, by life-cycle stage, over eight technology nodes 59
Fig. 4.7 Energy use per wafer by facility and process equipment,
         over eight technology nodes 60
Fig. 4.8 Life-cycle energy use per computational power 61
Fig. 4.9 GWP per computational power....................................... 61
Fig. 4.10 Energy use per die by life-cycle phase, 350 nm node 62
Fig. 4.11 GWP of emissions per die by life-cycle phase, 350 nm node....... 63
Fig. 4.12 Energy use per die by life-cycle stage, 45 nm node.
         Note: log scale attenuates appearance of uncertainty 63

Fig. 4.13 GWP of emissions per die by life-cycle phase, 45 nm
           node. Note: log scale attenuates appearance of uncertainty   64

Fig. 5.1   Life-cycle stages with data source types   71
Fig. 5.2   Global warming potential per die by life-cycle stage,
           over 7 technology nodes   77
Fig. 5.3   Water use per die, by life-cycle stage, over 7 technology
           generations   78
Fig. 5.4   Smog formation per die by technology node   79
Fig. 5.5   Acidification per die by technology node...........................   79
Fig. 5.6   Eutrophication via air emissions per die by technology node   80
Fig. 5.7   EPA criteria human health impacts per die by
           technology node...   80
Fig. 5.8   Human non-cancer health impacts per die by technology node   81
Fig. 5.9   Human carcinogenic impacts per die by technology node   82
Fig. 5.10 Ecotoxicity per die by technology node   82
Fig. 5.11 Worst-case eutrophication impacts (via water) per die
           by technology node   83

Fig. 6.1   Primary energy consumption per operating system
           equivalent memory capacity (MJ/GB), over five
           technology nodes...   89
Fig. 6.2   Environmental impacts due to air emissions per GB,
           over five technology nodes   90
Fig. 6.3   Human health impacts due to air emissions per GB,
           over five technology nodes   91
Fig. 6.4   GWP per GB memory capacity, by life-cycle stage,
           over five technology nodes   92
Fig. 6.5   Water consumption per GB memory capacity, by
           life-cycle stage, over the five nodes   93

Fig. 7.1   Primary energy consumption per OS memory capacity
           (MJ/OS), over five technology nodes   99
Fig. 7.2   Water consumption per OS memory capacity, by
           life-cycle stage, over five technology nodes 100
Fig. 7.3   GWP per OS memory capacity, by life-cycle stage,
           over five technology nodes 100
Fig. 7.4   Smog formation per OS by technology node 101
Fig. 7.5   Acidification per OS by technology node........................... 101
Fig. 7.6   Eutrophication via air emissions per OS by technology node 102
Fig. 7.7   Ecotoxicity per OS by technology node 102
Fig. 7.8   Criteria human health impacts per OS by technology node......... 103
Fig. 7.9   Carcinogenicity per OS by technology node 103
Fig. 7.10 Non-cancer human health impacts per OS by
           technology node... 104

Fig. 7.11   Comparison with Williams: Energy use at each
            life-cycle stage 105
Fig. 7.12   Comparison with Williams: Mass of process chemicals
            and gases 106

List of Tables

Table 2.1 Global Warming Potential (GWP) values for some
PFCs used in semiconductor manufacturing [63] 18

Table 2.2 IPCC Tier 2c standard values [62] 19

Table 3.1 Semiconductor products and their dominant device
types, by market volume 32

Table 4.1 Summary of data sources 38

Table 4.2 Chemical LCA data sources, part 1: process data 42

Table 4.3 Chemical LCA data sources, part 2: process data 43

Table 4.4 Chemical LCA data sources, part 3: EIO-LCA data.............. 43

Table 4.5 Chemical LCA data sources, part 3: process-based
common value [66] 44

Table 4.6 Energy intensity of silicon production 45

Table 4.7 Global warming intensity of Santa Clara water 45

Table 4.8 Cost of Fab infrastructure and equipment 47

Table 4.9 GWP intensity of electricity 48

Table 4.10 Electricity generation by type, world average vs. California 48

Table 4.11 Summary of process changes for each technology
generation, part 1 50

Table 4.12 Summary of process changes for each technology
generation, part 2 51

Table 4.13 Facility system changes by technology node 53

Table 4.14 Yields and chip sizes for each technology generation 54

Table 4.15 GWP intensity of transportation 55

Table 4.16 Transportation energy and CO_2 emissions by
technology node 56

Table 4.17 Use phase power by ITRS technology node 56

Table 4.18 Computational performance 57

Table 4.19 Data quality assessment 67

Table 5.1   POU abatement 75
Table 5.2   Use phase power by technology node 76
Table 5.3   VOCs lacking POF impact metrics 84
Table 5.4   Chemicals lacking toxicity impact metrics...................... 85

Table 6.1   Full life-cycle of 96 GB flash memory vs. HDD
             transportation and use 94

Table 7.1   Year and OS memory requirements for each
             technology node 98
Table 7.2   Chip size and power demand for each technology node........... 98

Table A.1   Energy consumption per die by life-cycle stage 113
Table A.2   Water consumption per die by life-cycle stage.................... 113
Table A.3   GWP per die by life-cycle stage 114
Table A.4   CMOS manufacturing inventory per wafer, 350 nm node......... 114
Table A.5   CMOS manufacturing inventory per wafer, 350 nm node......... 115
Table A.6   CMOS manufacturing inventory per wafer, 250 nm node......... 115
Table A.7   CMOS manufacturing inventory per wafer, 250 nm node......... 116
Table A.8   CMOS manufacturing inventory per wafer, 180 nm node......... 117
Table A.9   CMOS manufacturing inventory per wafer, 180 nm node......... 117
Table A.10   CMOS manufacturing inventory per wafer, 130 nm node......... 118
Table A.11   CMOS manufacturing inventory per wafer, 130 nm node......... 119
Table A.12   CMOS manufacturing inventory per wafer, 90 nm node 119
Table A.13   CMOS manufacturing inventory per wafer, 90 nm node 120
Table A.14   CMOS manufacturing inventory per wafer, 65 nm node 121
Table A.15   CMOS manufacturing inventory per wafer, 65 nm node 121
Table A.16   CMOS manufacturing inventory per wafer, 45 nm node 122
Table A.17   CMOS manufacturing inventory per wafer, 45 nm node 123
Table A.18   CMOS manufacturing inventory per wafer, 32 nm node 124
Table A.19   CMOS manufacturing inventory per wafer, 32 nm node 124
Table A.20   CMOS Manufacturing emissions per wafer, 350 nm node 125
Table A.21   CMOS Manufacturing emissions per wafer, 350 nm node 126
Table A.22   CMOS Manufacturing emissions per wafer, 250 nm node 127
Table A.23   CMOS Manufacturing emissions per wafer, 250 nm node 128
Table A.24   CMOS Manufacturing emissions per wafer, 180 nm node 129
Table A.25   CMOS Manufacturing emissions per wafer, 180 nm node 130
Table A.26   CMOS Manufacturing emissions per wafer, 130 nm node 131
Table A.27   CMOS Manufacturing emissions per wafer, 130 nm node 132
Table A.28   CMOS Manufacturing emissions per wafer, 90 nm node 133
Table A.29   CMOS Manufacturing emissions per wafer, 90 nm node 134
Table A.30   CMOS Manufacturing emissions per wafer, 65 nm node 135
Table A.31   CMOS Manufacturing emissions per wafer, 65 nm node 136
Table A.32   CMOS Manufacturing emissions per wafer, 45 nm node 137
Table A.33   CMOS Manufacturing emissions per wafer, 45 nm node 138

Table A.34 CMOS Manufacturing emissions per wafer, 32 nm node 139
Table A.35 CMOS Manufacturing emissions per wafer, 32 nm node 140
Table A.36 Life cycle impacts per die, 32 nm node 142
Table A.37 Life cycle impacts per die, 45 nm node 143
Table A.38 Life cycle impacts per die, 65 nm node 144
Table A.39 Life cycle impacts per die, 90 nm node 145
Table A.40 Life cycle impacts per die, 130 nm node 146
Table A.41 Life cycle impacts per die, 180 nm node 147
Table A.42 Life cycle impacts per die, 250 nm node 148
Table A.43 Life cycle impacts per die, 350 nm node 149

Table B.1   Flash memory manufacturing inventory per wafer,
            150 nm node 151
Table B.2   Flash memory manufacturing inventory per wafer,
            150 nm node 152
Table B.3   Flash memory manufacturing inventory per wafer,
            120 nm node 153
Table B.4   Flash memory manufacturing inventory per wafer,
            120 nm node 153
Table B.5   Flash memory manufacturing inventory per wafer,
            90 nm node 154
Table B.6   Flash memory manufacturing inventory per wafer,
            90 nm node 155
Table B.7   Flash memory manufacturing inventory per wafer,
            65 nm node 156
Table B.8   Flash memory manufacturing inventory per wafer,
            65 nm node 156
Table B.9   Flash memory manufacturing inventory per wafer,
            45 nm node 157
Table B.10 Flash memory manufacturing inventory per wafer,
            45 nm node 158
Table B.11 Flash memory manufacturing emissions per wafer,
            150 nm node 159
Table B.12 Flash memory manufacturing emissions per wafer,
            150 nm node 160
Table B.13 Flash memory manufacturing emissions per wafer,
            120 nm node 161
Table B.14 Flash memory manufacturing emissions per wafer,
            120 nm node 162
Table B.15 Flash memory manufacturing emissions per wafer,
            90 nm node 163
Table B.16 Flash memory manufacturing emissions per wafer,
            90 nm node 164
Table B.17 Flash memory manufacturing emissions per wafer,
            65 nm node 165

Table B.18 Flash memory manufacturing emissions per wafer,
            65 nm node 166
Table B.19 Flash memory manufacturing emissions per wafer,
            45 nm node 167
Table B.20 Flash memory manufacturing emissions per wafer,
            45 nm node 168
Table B.21 Flash life cycle impacts per wafer, 45 nm node.................... 169
Table B.22 Flash life cycle impacts per wafer, 45 nm node.................... 170
Table B.23 Flash life cycle impacts per wafer, 65 nm node.................... 171
Table B.24 Flash life cycle impacts per wafer, 65 nm node.................... 172
Table B.25 Flash life cycle impacts per wafer, 90 nm node.................... 173
Table B.26 Flash life cycle impacts per wafer, 90 nm node.................... 174
Table B.27 Flash life cycle impacts per wafer, 120 nm node 175
Table B.28 Flash life cycle impacts per wafer, 120 nm node 176
Table B.29 Flash life cycle impacts per wafer, 150 nm node 177
Table B.30 Flash life cycle impacts per wafer, 150 nm node 178

Table C.1   DRAM manufacturing inventory per wafer, 250 nm node 179
Table C.2   DRAM manufacturing inventory per wafer, 250 nm node 180
Table C.3   DRAM manufacturing inventory per wafer, 180 nm node 181
Table C.4   DRAM manufacturing inventory per wafer, 180 nm node 181
Table C.5   DRAM manufacturing inventory per wafer, 130 nm node 182
Table C.6   DRAM manufacturing inventory per wafer, 130 nm node 183
Table C.7   DRAM manufacturing inventory per wafer, 90 nm node.......... 184
Table C.8   DRAM manufacturing inventory per wafer, 90 nm node.......... 184
Table C.9   DRAM manufacturing inventory per wafer, 70 nm node.......... 185
Table C.10 DRAM manufacturing inventory per wafer, 70 nm node.......... 186
Table C.11 DRAM manufacturing inventory per wafer, 57 nm node.......... 187
Table C.12 DRAM manufacturing inventory per wafer, 57 nm node.......... 187
Table C.13 DRAM manufacturing emissions per wafer, 250 nm node........ 188
Table C.14 DRAM manufacturing emissions per wafer, 250 nm node........ 189
Table C.15 DRAM manufacturing emissions per wafer, 180 nm node........ 190
Table C.16 DRAM manufacturing emissions per wafer, 180 nm node........ 191
Table C.17 DRAM manufacturing emissions per wafer, 130 nm node........ 192
Table C.18 DRAM manufacturing emissions per wafer, 130 nm node........ 193
Table C.19 DRAM manufacturing emissions per wafer, 90 nm node 194
Table C.20 DRAM manufacturing emissions per wafer, 90 nm node 195
Table C.21 DRAM manufacturing emissions per wafer, 70 nm node 196
Table C.22 DRAM manufacturing emissions per wafer, 70 nm node 197
Table C.23 DRAM manufacturing emissions per wafer, 57 nm node 198
Table C.24 DRAM manufacturing emissions per wafer, 57 nm node 199
Table C.25 DRAM life cycle impacts per wafer, 57 nm node 201
Table C.26 DRAM life cycle impacts per wafer, 57 nm node 202
Table C.27 DRAM life cycle impacts per wafer, 70 nm node 203
Table C.28 DRAM life cycle impacts per wafer, 70 nm node 204

Table C.29   DRAM life cycle impacts per wafer, 90 nm node 205
Table C.30   DRAM life cycle impacts per wafer, 90 nm node 206
Table C.31   DRAM life cycle impacts per wafer, 130 nm node 207
Table C.32   DRAM life cycle impacts per wafer, 130 nm node 208
Table C.33   DRAM life cycle impacts per wafer, 180 nm node 209
Table C.34   DRAM life cycle impacts per wafer, 180 nm node 210
Table C.35   DRAM life cycle impacts per wafer, 250 nm node 211
Table C.36   DRAM life cycle impacts per wafer, 250 nm node 212

Acronyms and Abbreviations

Acronyms

CMOS	Complementary metal oxide semiconductor
CMP	Chemical mechanical polishing
DRAM	Dynamic random access memory
EOL	End-of-life
GWP	Global warming potential
ICT	Information and communication technologies
NAICS	North American Industry Classification System
PFC	Per-fluorinated compound

Chemical Abbreviations

BTBAS	Bis tertiary-butylamino silane
DEA	Diethyl-amine
HMDS	Hexa-methyl disilizane
IPA	Isopropyl alcohol
OMCTS	Octamethylcyclotetrasiloxane
PDMAT	Pentakis(dimethylamido)tantalum
PGMEA	Propylene glycol monomethyl-ether acetate
TDEAH	Tetrakis(diethylamino)hafnium
TDMAS	Tris(dimethylamino)silane
TMAH	Tetramethyl ammonium hydroxide

Chapter 1
Introduction

Information and communication technology (ICT) has the potential to reduce the impact of human activities on the environment. In order to fully understand the environmental benefits of ICT, the life-cycle impacts of computer systems must be compared with those of the products and services they replace. The questions of whether reading news on a handheld device rather than newspaper, or purchasing books from an online retailer instead of from a bookstore reduces environmental impact are two examples of this sort of comparison in the recent literature [65, 100, 123]. While, initially, the replacement of traditional products such as newspapers by a small fractional increase in the use of a handheld mobile device seems a winning environmental trade-off, there has been increasing concern over the large energy demands of the internet infrastructure, with data center energy demand in the U.S. reaching 1.5% of the national total in 2006 and estimates of 2011 demand surpassing 10 billion kWh [36].

Among the numerous parts which compose the IT infrastructure, semiconductor chips are among the most resource-intensive to produce as well as the most difficult to characterize for the purposes of life-cycle assessment (LCA). While it may be possible to estimate the environmental impacts of a cable or plastic computer housing knowing only their masses and material types, the impacts associated with a semiconductor chip are not represented well by the substance of the device itself. While a logic chip may weigh only a few grams, the chemicals and water required to produce it weigh many kilograms. In many LCA studies of electronics, the lack of LCA data for computer chips has been noted as an important topic for future work and the need for a more detailed and transparent life-cycle inventory for semiconductor products has been noted previously [12, 79, 93, 94, 125, 130].

The purpose of this book is to inform and facilitate life-cycle assessment (LCA) of semiconductor devices and electronics in general. Public policy and research which is informed by representative models of the life-cycle environmental and human health impacts of electronic products can more effectively govern their production, use and end of life.

S.B. Boyd, *Life-Cycle Assessment of Semiconductors*, DOI 10.1007/978-1-4419-9988-7_1,   1
© Springer Science+Business Media, LLC 2012

LCA is a powerful environmental management approach which has developed into the common method of accounting for the environmental impacts of products. LCA requires the analyst to account for a broad set of environmental and human health impacts which occur throughout the production, use and end-of-life of the product. Due to its required scope, LCA prevents the offloading or externalization of environmental burdens from one life-cycle stage or impact category to another, which may otherwise occur when a more limited view is taken in the environmental management of a product.

This book describes the available analytical techniques for LCA of semiconductors, as well as a set of life cycle inventory (LCI) data for several semiconductor device types, which may be adapted for use in LCA. Using the methods and data presented herein, LCA practitioners among policy-makers and the public can better understand and manage the impacts of electronic products and services.

1.1 Semiconductor Life-Cycle Environmental Impacts

Semiconductor production is highly resource intensive and generates a wide variety of emissions, some of which have global effects. The processes used to manufacture semiconductors emit several major classes of pollutants, including global warming gases (e.g. CF_4, NF_3, C_4F_8), ground level ozone-forming volatile organics (e.g. isopropyl alcohol, formaldehyde), hazardous pollutants (e.g. arsenic, fluorine) and flammable materials (e.g. silane, phosphine). Semiconductor fabrication facilities also consume large volumes of water and energy, and the high purity chemicals used in production are highly refined and thus have high "embodied energy". The upstream environmental effects due to chemicals manufacturing, as well as fabrication facility (fab) infrastructure and equipment, represent significant components of the environmental impact profile of semiconductor manufacturing. The use phase of semiconductor devices results in indirect environmental and human health impacts resulting from energy-related emissions which, in the case of logic devices, has been shown to dominate impacts over the product life-cycle. The end of life of a semiconductor chip results in lead emissions to soil if there is lead present in the chip's leadframe solder. After 2006, the EU's Restriction on Hazardous Substances, commonly known as RoHS, banned the use of lead in electronics and most major manufacturers with global markets switched to lead-free solders worldwide to comply with this regulation. Other effects from end-of-life disposal of semiconductor devices, such as emissions from incineration, may occur in other end-of-life scenarios.

The semiconductor manufacturing also poses hazards to workers and technicians in the exceptionally large variety of chemistries which it employs. Wafer processing involves a number of different acidic (the hydrofluoric and sulfuric acids used in wafer cleans), basic (wafer clean steps including ammonia), oxidizing (wafer cleans using peroxide) and other highly reactive chemistries (fluorine used in etching), as well as compounds which are extremely toxic (arsine and phosphine used in implant). The equipment used to administer these reactions must be designed to protect the manufacturing personnel, following safety rules outlined by government agencies such as OSHA and standards (e.g. SEMI S2) developed within industry groups such as Sematech.

As all mainstream semiconductor manufacturing equipment currently sold and used follows these regulations, the direct human health impacts and risks within the fab have been nearly eliminated in normal operation. (Though, hazards still exist in cases of catastrophic breakdown, fire or earthquake.) Once these chemicals leave the equipment, they must be further handled and neutralized by the POU and facility abatement systems, in a safe and efficient way. While the guidelines and standards for equipment safety are enough to thoroughly guide and ensure the design of safe equipment, the design and operation of facility abatement is a much more complex undertaking. The abatement and neutralization of emissions is not as predictably efficient or controlled as the reaction of chemicals within the process equipment in part because the processes used to neutralize emissions to the extent necessary to make them safe for release into the environment do not need to be as precise as those used within the process chamber. Additionally, within the facility abatement systems (the house gaseous waste, fluorine abatement and acid waste neutralization systems), the chemistry of the combined emissions of the many processes running on site can be unpredictable. Facility abatement systems are designed to continuously measure the incoming waste stream, and adjust the neutralization chemistry accordingly. Nevertheless, neutralization of an unpredictable waste stream cannot be as efficient or controlled as that of a known waste stream.

When facility abatement systems are not operating ideally, or were not originally designed or built to sufficiently handle the current waste streams entering them, a variety of environmental impacts can result. For example, the "house scrubber" (facility gaseous abatement system) may be accepting significant concentrations of gaseous fluorine (F_2), either because no point-of-use abatement is set up on plasma etching equipment, or because POU systems are not sufficiently scrubbing the F_2 gas. This gaseous fluorine will react with water to a small extent to form OF_2, a reactive and highly toxic compound [110]. Another product of the reaction of fluorine with water is HF. When fluorinated compounds are effectively abated from processes at POU, the resulting liquid HF is sent to a fluorine waste treatment system which is separate from the house acid waste neutralization system. Any HF captured in the house scrubber system could not be effectively treated before being released into the environment, as it would already be mixed in with the larger volume of non-hazardous waste. Ineffective abatement of fluorine, and the consequent release of reactive fluorine species into the environment could result in human health and ecological impacts. The unknowns associated with underperforming or poorly designed abatement and waste handling systems are a difficult aspect of semiconductor LCI to effectively quantify.

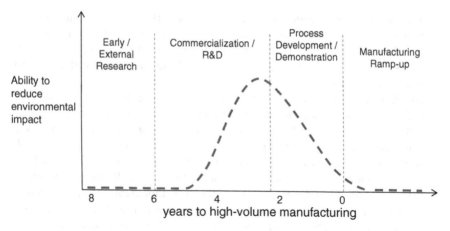

Fig. 1.1 Development timeline for an advanced semiconductor product

1.2   Challenges in LCA of Semiconductors

There are many challenges in application of LCA to integrated circuits. The relatively short development cycle and rapid technology change in design and manufacturing make assessment of existing chips obsolete in just a few years [51]. (A generalized development timeline for semiconductor products is illustrated in Fig. 1.1.)

IC manufacturing is also highly complex, involving hundreds of chemicals and dozens of different types of process steps, combined in a process flow which typically includes hundreds of steps. Intellectual property issues also plague data collection, as semiconductor manufacturers consider process recipes to be their most valuable intellectual assets, chemical suppliers often keep the formulations of process chemicals trade secrets, and emissions measurement is time consuming and requires specialized equipment and methods. Nevertheless, significant progress has been made on the topic of semiconductor LCA, as demonstrated in the following section, which relates a chronological timeline of publicly-available research on semiconductor LCA.

1.3   Timeline of Semiconductor LCA in Practice

A widely-cited, definitive report, and evidently the earliest study of the life-cycle assessment of electronics is a document entitled "Environmental consciousness: a strategic competitiveness issue for the electronics and computer industry" [77]. This report which was commissioned by the EPA and developed with help from the Microelectronics and Computer Technology Corporation was commonly referred to as the "MCC study" or "MCC report". While no longer publicly available, the EPA summarizes the report's major findings as being the identification of wet chemistry processes, such as those used in wafer processing and printed wiring board (PWB) fabrication, as being of environmental importance due to their high chemical use as well as energy and water demands, as well as the need for collaborative research between industry, academia and the government to further address the environmental management of electronics manufacturing [67].

On the topic of life-cycle assessment of semiconductors in particular, one of the earliest publicly available, published works relating to semiconductor LCA can be found in a 1997 conference paper from NEC Corporation on comparative LCA of two personal computers (PCs): a laptop and a desktop [120]. The paper provides significant detail into the quantities and types of environmental impacts from the PC life-cycle but limited insight into the methodology behind the analysis. Although the authors of the paper were within a computer manufacturing company, they based the inventory only on a disassembly analysis; Rather than accounting for materials and energy used in computer assembly, they base the inventory only on the parts and subcomponents present in the final products. The life-cycle inventory (LCI) data for the electronic components, including the semiconductor chip, liquid crystal display and cables, were based on economic input-output data for Japan, which implies that direct emissions from semiconductor fabrication were not included. The paper reports quantitative life-cycle impacts for global warming potential, acidification, eutrophication and resource consumption, as well as the proportional impact of each component in the production stage. The results are detailed, and particularly useful in that each subcomponent's contribution to impacts are reported. However, there were many important parts of the inventory that were left out, including resource consumption and emissions from semiconductor fabrication.

The following year, researchers involved in the development of the Ganzheitliche Bilanzierung (GaBi) LCA software at University of Stuttgart presented a methodology paper on green design for electronics making the case for LCA as the best tool for the job [11]. The paper described some important aspects of semiconductor and electronics LCA, stating that feedback concerning the environmental consequences of production and use were necessary in the research and development stages in order to influence design, rather than after production ramp-up. Also, the paper pointed out that because of the complexity of electronic products, their dense supply chains and the distributed decision-making process behind their production, the development of an electronics LCA database would require intra-industry collaboration and information sharing and that the task was onerous, involving an "immense demand of information." The paper also pointed to the fact that the environmental impacts of production were not evident in the products themselves, as the toxic chemicals used in manufacturing do not end up in the final product. In the concluding statements, the author prioritizes future work, asserting that database development for generic chip production was of high priority.

A researcher in Japan used economic input-output techniques to estimate LCI data for several electronics components (including semiconductors, passive components and liquid crystal displays) [125] and came to some interesting conclusions. Industry-wide emissions data were taken from government sources in Japanese, and thus the accuracy and representativeness of the data are difficult to determine. The author notes that the data sets are missing some component materials and emissions data, and that direct greenhouse emissions from manufacturing sites are not included in the inventory. Ueno finds that impacts associated with chemical usage are large but difficult to estimate and concludes that LCI data is similar among all electronic components industries, which is a surprising assertion.

In 2001, a conference paper describing a life-cycle inventory (LCI) model for a semiconductor wafer was written by an academic with support of industry members at Motorola [102]. The purpose of the study was to investigate the most important environmental impacts of the fabrication facility. There was particular attention paid to the potential for toxic exposures within the fab. The analysis used ProTox, a toxicological screening tool to evaluate relative risks. Schischke described an equipment-centric inventory method whereby mass and energy flows are accounted for in modules specific to process types and facility infrastructure. However, the inventory inputs were collected by questionnaire from anonymous industry members and outputs were estimated as fractions of the input flows. In addition, allocation is addressed as an issue, meaning that there were several products being produced in the plant, and impacts had to be allocated among them. Concerning data quality, Schischke wrote:

"For production and infrastructure processes, few data are known exactly, but estimations of experts concerning a suitable allocation of mass flows leads to sensible and high quality results."

At the time, these expert opinions and process-level input-output estimations (estimating the emissions based on the process recipe) were the best semiconductor LCI data that had been compiled and reported publicly, and this report was an important step forward. As the purpose of the study was internal environmental management, only a limited summary of inventory and impact results were given, and no absolute impact results were shown by process, rather only the proportional contribution of each process module to each impact type. The study could therefore not be directly used by other LCA researchers seeking LCI data for semiconductors in the assessment of electronics.

At the same 2001 conference, another paper from ST Microelectronics reports a "gate-to-gate" life-cycle inventory (LCI) analysis for an 8 Mbit EPROM chip [119]. The form of the model is equipment-centric, rather than facility-based: data for the chemicals, facilities resources (e.g., ultra-pure water) and electricity demands for each process step were collected, and summed to represent the entire process flow for the EPROM device. The inventory of the masses of materials is reported, making the study more transparent than the Motorola study, and transportation from the front-end to the back-end facility was included, making the analysis more inclusive than the previous papers. Process and facility emissions, were not included in the LCI, however.

Several other descriptions of internal green design efforts by semiconductor industry members arose in trade magazines and conference proceedings [46,72,116] but there had been no peer-reviewed journal articles directly addressing LCA of semiconductor devices until 2002 when Eric Williams published an estimate of the energy and materials demands for a 32 MB DRAM chip in a paper entitled "the 1.7 Kilogram Microchip" [130]. The paper provided an abbreviated list of material inputs to semiconductor fabrication from an anonymous industry source, compared this with previous estimates and called for more accurate process-level LCI data for semiconductor chips. One of his main contentions was that the "upstream" energy used to make process chemicals would be very large due to the extremely high purity of semiconductor process chemicals and because the use of byproduct or "secondary" materials is much higher for semiconductors than for traditional products.

In 2003, Cynthia Murphy at the University of Texas at Austin described a plan to create parametric process models for semiconductor LCI [79]. Murphy measured power consumption and gas flows for several types of furnace-based thermal processes and provided an equation to estimate power and gas consumption based on oxide film thickness. The paper also provided power consumption data for photolithography. These LCI process data and the concept of parametric LCI models were useful and innovative. Unfortunately there was no further work by Murphy on semiconductor LCI and without LCI models for all of the semiconductors processes (i.e., ion implant, chemical vapor deposition, etc.) it would not be possible to model a complete device. Although the idea of parametric LCI modeling is attractive, in practice these models cannot predict resource consumption outside of the range of measured data. For example, the equation presented by Murphy in the paper for thermal processes does not give a good prediction of process gas and power consumption outside of the measured scope of the model (e.g. for oxide films thinner than the thinnest film in the model).

Other researchers besides Murphy have measured and reported the environmental-relevant data for specific semiconductor manufacturing process steps in the context of LCA. For example, Paul Blowers measured LCI data for supercritical CO_2 as a replacement for ultra-pure water [13]. There has also been a particular focus on solder and packaging technologies, due to the exceptional hazard of lead, a potent neurotoxin, in solder [6, 33, 47, 80] as well as the relative environmental impacts of replacement PFCs and advancements in plasma technologies for etching [97].

In 2004, a life-cycle inventory for two generations of processor chips developed at Intel was presented in a conference paper [134]. In this study, the emissions and energy associated with each chip were based on facility-level data, i.e., the power, water, emissions and materials consumption for an entire fab were divided by production output to determine the impacts of a single device. The comparison of a 60 nm Pentium Pro processor with a 130 nm Pentium 4 revealed that the switch from 200 to 300 mm wafers resulted in significant reductions in emissions per chip. Comparison among life-cycle stages also showed that use phase dominates life-cycle energy use. While these results were useful, the study methodology was opaque. The life-cycle analysis apparently included only wafer production and chip use, and not back end operations, transportation, end of life or any upstream impacts. It is not clear how emissions were measured or estimated and emissions were presented by category rather than by chemical, which suggests that they are estimated from permits rather than measurements. The study reported ultra-pure water rather than actual water consumption, which is not a useful proxy for actual water use and may be misleading to some audiences.

Xiaoying Zhou, in 2007, completed a book at UC Davis entitled "Life-cycle thinking and assessment tools on environmentally benign electronics." The work includes the development of an integrated impact assessment and weighting methodology and two case studies of green electronics: the introduction of lead-free solder and remanufacturing of a cellular phone [137].

The work presented in this book is the latest output of a long-running research partnership between University of California at Berkeley and the semiconductor equipment supplier Applied Materials. This collaboration initially focused on addressing the problem of externalization of impacts through the integration of human health and environmental impact metrics with economic costs in decision-support systems. In 2000, Thurwachter developed an environmental cost of ownership model termed EnV (environmental value), which assessed operating costs and emissions and included a novel multi-criteria health hazard scoring system [121, 122]. This tool was a form of enhanced life-cycle inventory modeling which was specifically designed to inform semiconductor manufacturing equipment designers and process engineers of human health and environmental risks. At that stage, the evaluation of environmental impacts was not yet directed towards LCA of a semiconductor product. As LCA became a more dominant theme in green design and manufacturing, the models developed in the research partnership become more directed towards life-cycle thinking. Krishnan, in 2003, extended the EnV model by integrating Monte Carlo simulation and applied it to a variety of processes and equipment. The updated model, named EnV-S (environmental value systems), was used to determine environmental and economic cost of chemical mechanical planarization (CMP) process tools, as well as numerous fluorine and per-fluoro-compound (PFC) processes and abatement systems [70,72]. Krishnan also provided intellectual contribution to the development of the first iteration of the life-cycle inventory models presented in the current work [71].

Some valuable conclusions to be made in review of the published literature on LCI of semiconductors is that impacts resulting from electricity consumption in the use phase, manufacturing energy use (including both electricity and other energy sources used in wafer fabrication and chip packaging) and process emissions in wafer fabrication are the three aspects of the life-cycle which stand out among the rest. Nearly all studies which include the use phase find it to consume the most energy or produce the largest quantity of impacts resulting from air emissions among the life-cycle stages [15a, 119, 130, 134]. Those studies which include manufacturing, but expand the boundary of analysis to include at least some other life-cycle stages, such as transportation and the production of some process materials, find that electricity use for manufacturing is the next largest source of environmental impacts after the use phase [102, 119, 130, 134]. In those studies which include manufacturing, use and additional life-cycle phases in which PFC emissions are included in the inventory for manufacturing, find that PFCs contribute a large fraction of GWP impacts, in some cases approaching the magnitude of GWP impacts resulting from manufacturing electricity [15a, 15b, 30]. While some of the studies available in the literature reflect older technologies, the dominance of these three sources of environmental impacts reflects the current manufacturing setting as well, as these conclusions are also supported by the most recent report by Intel to the Carbon Disclosure Project [61]. LCA of semiconductors should be approached with an understanding of the relative importance of these stages (use phase, wafer fabrication, and chip assembly/packaging) of the life-cycle.

This literature timeline has focused on the application of LCA to semiconductors, though LCA has also been used to study many other aspects of electronics production besides semiconductors. Some of the most comprehensive and reliable work on LCA of other electronic components has been conducted by the US EPA. The first study mentioned in this review, the MCC study, was an early example of this research effort which was initiated by the EPA's Office of Pollution Prevention and Toxics Design for Environment (DfE) Program. The MCC study's identification of PWB processing as an environmental hot-spot in the computer life-cycle influenced the EPA's research strategy concerning environmental management of computer production in the late 1990s, during which time the agency focused particular research attention on the PWB industry, publishing a number of industry surveys and case studies. The EPA's DfE program continued to output valuable research, releasing a comprehensive LCA of computer monitors in 2001 [34] and LCA of leaded and lead-free solders in 2005 [35]. Though the agency has not used LCA publicly as a key tool for evaluation of semiconductors or for environmental management of semiconductor production, the EPA's LCA studies on solder and displays as well as the agency's continuing research in this area are valuable resources for practitioners of electronics LCA.

1.4   LCA of Information Technology

The use of electronics as a replacement for traditional media is a promising strategy for environmental impact reduction of various products, services and human activities. Consumer electronics can convey reading materials without printed media, improved telecommunication may be used to reduce individual travel and electronic controls can enhance the efficiency of existing technologies in a variety of applications. Notable early studies of the prospective environmental advantages or tradeoffs include Toffel's comparison of printed and electronic news sources [123] and Plepys' investigations into the environmental implications of ICT-enabled dematerialization [91, 92, 95]. More recently, there have been institutional research efforts to address questions about the potential environmental benefits of ICT, such as the Global e-Sustainability Initiative's Smart 2020 Report [48, 49], and the use of ICT to reach climate change mitigation goals, most notably research directed by the European Commission [31, 32].

Current, detailed and technically specific models of the impacts of semiconductor products are a requirement for an accurate comparison between the life-cycle impacts of existing practices or technologies and their electronic replacements or enhancements. In nearly every publicly reported study on the environmental benefit or damage of information technology (IT), there has been an expressed or implied need for more detailed and accessible LCA data for semiconductor devices. In this book, life-cycle inventory models for many generations of commonly used logic and memory components are presented with the purpose of meeting this need and enabling more effective environmental management in the use of electronic products or IT in the replacement or modification of traditional services.

1.5 Overview of this Book

This book presents an overview of methods used in semiconductor LCA, as well as a number of case studies which may be read for their value as examples of semiconductor LCA, or consulted for LCI data for use in LCA of electronics. These LCA case studies are multi-generational process-level LCA studies of common semiconductor products: CMOS logic, flash memory and DRAM, which are reported in a way that facilities downstream use.

The LCA performed within this book are based on equipment or process-level, rather than facility-level, data, which allowed more rapid development of multi-generational LCA, overcoming the problem of LCA obsolescence. Process-level LCA of semiconductor manufacturing allows anticipative models of manufacturing production because if the process flow is known, a new device can be modeled pre-production using previously measured process data, and without facility data collection. Process-specific analysis also allows more accurate accounting of the upstream impacts associated with high-purity materials. While facility LCI models typically report chemical use data based on chemical purchases, a process-based model can determine more accurately the volume of chemicals used per finished wafer.

The data used for each individual wafer processing step is from the Supporting Information of a paper reported by Applied Materials [71]. These process step LCIs were developed from a database of emissions measurements and power tests which had been populated by the environmental health and safety and emissions testing groups Applied Materials, over a period of more than a decade. In the LCA presented in this book, these previously-reported individual process steps were combined into process flows which were developed by the author through consultation with the literature and industry documentation, as well as expert advice from academics and members of industry (Fig. 1.2).

The process flow LCI were then linked with facility models, which represent all of the main facility utilities (ultra-pure water purification, clean-room air supply, etc.) with generationally-specific energy consumption factors, fab construction costs representative of the year of study, as well as facility abatement models, which

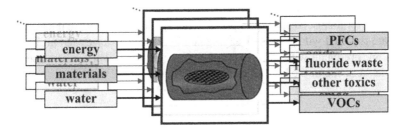

Fig. 1.2 Compilation of individual process models into process flows

accept all post-abatement emissions flows. The compiled material demands for the LCI at each generation are also linked on the upstream side to the mass-normalized impacts for chemicals production. The LCIs reported in this book differ from those reported in previous papers by the author [15a, 15b, 15c], as they have been updated in collaboration with David Bol [14]. Beyond the life-cycle stage of semiconductor wafer fabrication, LCI data is taken from academic literature as well as economic input-output modeling, where no process data was available.

The transparency of the methodology and background data used in the LCI and LCA studies presented in this book allow the results to be more easily adapted for use in LCA of electronics. Although several industry members have published papers which present LCA results for their products, the methodology and data sources used in those analyses are in some cases not fully described. In such cases, data quality and the uncertainty associated with the results are difficult to assess. Throughout this book, the sources of data and the uncertainty associated with data are reported and the methodology is described in detail, such that the analysis is reproducible and adaptable. At each stage the uncertainty associated with each data source is reported, propagated through the model and accounted for in the upper bound, lower bound, and base case results.

The LCA studies presented in this book are also useful to LCA practitioners because they assess a broader boundary and report results by life-cycle stage, to allow boundary redefinition in downstream application. In many previous studies, portions of the life-cycle, such as the upstream impacts related to chemicals, have been excluded from the scope, as appropriate for each study's particular goal. Because the goal of this study is to present LCI for use by LCA practitioners, rather than to investigate one stage in detail, the boundary need not be limited. The studies presented in the book thus attempt to evaluate all life-cycle stages, to the extent practicable, including even those stages of lesser importance such as municipal water treatment and delivery. By including all stages, these LCA of semiconductors allow some of the lingering questions concerning the possible contribution of omitted aspects of the life-cycle (e.g., impacts associated with fab construction, equipment, chemicals and water) to be addressed. The author does not attempt to present a concrete resolution on whether to include or exclude certain stages, such as upstream chemicals production and the construction of the fab infrastructure and equipment. Individual LCA practitioners must make these judgments for themselves, as appropriate to the application. Because the LCI in this book are reported with results disaggregated by life-cycle stage, and all life-cycle stages are presented with uncertainty bounds or best and worst cases, users may redefine the scope of the LCI for their own purposes.

Chapter 2
Semiconductor LCI Methods

Sources of environmental impacts in IC production and use include emissions from electricity generation and fuel incineration as well as direct process emissions at manufacturing sites. Looking upstream, environmental impacts also occur due to process heating, electricity generation, and direct emissions in several additional life-cycle stages, including production and purification of silicon, infrastructure construction, equipment manufacturing and chemicals production. Transportation also contributes to impacts as a component of these upstream life-cycle stages, and as a stage of its own, when wafers are conveyed from the wafer fabrication facility to the assembly plant and chips are transported to the point of use. Figure 2.1 depicts the sequence of some of the key life-cycle stages for a semiconductor product, as well as some environmentally significant ancillary stages.

A variety of methods have been used to evaluate the energy use, global warming impact or other environmental effects resulting from the production and use of semi-conductors. In this chapter, a summary of the techniques available for evaluation or estimation of semiconductor LCI data, including upstream processes (chemicals, facility and equipment production) will be presented. Particular attention will be paid to LCI of wafer fabrication, and the customization of existing integrated circuit LCI data for downstream use, in LCA of electronics.

2.1 LCI of Wafer Fabrication

In order to develop an LCI for the life-cycle stage of wafer fabrication, power, water and materials use, and facility emissions should be accounted for. An LCI may be developed by develop measuring or taking data on site, using established estimation methods such as those developed by governments, or by adapting existing LCI data for wafer production of an equivalent IC to the application. In this section, these three methods will be described in detail.

S.B. Boyd, *Life-Cycle Assessment of Semiconductors*, DOI 10.1007/978-1-4419-9988-7_2,   13
© Springer Science+Business Media, LLC 2012

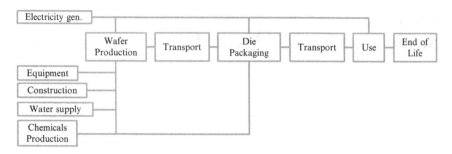

Fig. 2.1 Life-cycle stages of an integrated circuit

2.1.1 Mass and Energy Flow Modeling

Life-cycle inventory data for a semiconductor device may be developed using measured or estimated data taken at the facility or equipment level (aka, process level). An example of data taken at the facility level includes the total annual power usage by the fab, which could be determined by reading the building power meter. Data taken at the equipment level would include chemical usage in a particular process step, which would be metered at the inlet of the manufacturing equipment. The advantage of facility data is that is more easily obtained; however, if the facility is used to produce different types of devices, the facility's resource uses and emissions must be allocated among those products. The choice of allocation method (economic, per unit, by mass) is essentially a qualitative decision and allocation introduces uncertainty into the LCI data. Equipment-level data, because it involves smaller mass flows, is more precise than facility data. In addition, it does not entail allocation when collected in a multi-product facility. However, basing an LCI entirely on process-level data requires testing power consumption and measuring chemical use and emissions for each and every step in the process flow of a device, which is usually not feasible. In addition, equipment data alone does not represent all of the energy used in production, as the facility systems which provide the cleanroom environment, process cooling water, ultrapure water and manage emissions must also be accounted for in the LCI. When possible, collection of data at both the facility level and equipment level allows error checking. Power may be measured at the equipment level, for example, and chemical usage may be estimated using data from facility purchases. When both equipment-level and facility-level data are available, facility data and results from a process model based on equipment data may be cross checked against one another. Once data for both the resources uses and emissions of the production process are known, a mass balance of the system may also be used as an error checking tool.

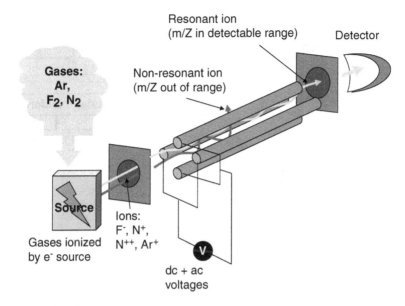

Fig. 2.2 Quadrupole mass spectrometry

2.1.1.1 Semiconductor Process Emissions Measurement

Mass emissions can be measured at the manufacturing tool chamber inlet and outlet using a combination of quadrupole mass spectrometry (QMS), Fourier transform infra-red (FT-IR) spectroscopy, and chemiluminescence-based fluorine gas analysis [109]. The combination of analytical techniques required is dependent on the possible range of chemicals present.

QMS is a popular mass measurement technique in which the molecular or atomic masses of species present in a gas medium are determined through ionization of the gas and subsequent electrical resonance and acceleration of the charged ions. As shown in Fig. 2.2, the four electrical poles carry ions towards the detector, but at any given moment, the generated charge only carries those ions with a particular mass to charge ratio. The poles sweep through a spectrum of charge, delivering a corresponding mass spectrum to the detector.

QMS is not sufficient to detect all materials present when multiple ionized species with common mass to charge ratios are present.

FT-IR spectroscopy is a powerful analytical technique which is used to identify and quantify the chemicals present in a medium based on their signature absorbance of the IR spectrum. A diagram of an FT-IR analyzer is shown in Fig. 2.3. A beam of collimated, polychromatic infrared light is split, such that one half is sent to a moving mirror, and the other half is directed toward a static mirror. These beams reflect back, are transmitted through the sample and are recorded by the detector. The recording taken over each full length of travel by the moving mirror is called an

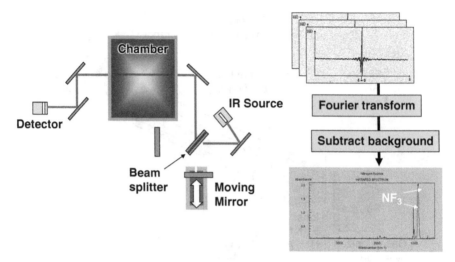

Fig. 2.3 Fourier transform infrared spectroscopy

interferogram. The set of all interferograms across the spectrum are transformed (via Fourier transform) from the physical length domain into the wavenumber domain, the background spectrum is subtracted, and the resulting spectrum can be used to identify the species present.

Water and carbon dioxide, which are both nearly ubiquitous and difficult to remove from vacuum environments, interfere with sizeable ranges of the IR spectrum. When the background spectrum, which contains the spectra representing the water vapor, carbon dioxide and other gases present in the chamber before the measurement is taken, is subtracted out, the differences between the background scan and the actual concentration of water and CO_2 present during the measurement scan result in noise and information loss over those spectral areas. This limitation, inherent in the technology, creates difficulties in the interpretation of species over the water and CO_2 spectral areas. QMS may be used in complement to FT-IR to overcome these informational losses, as QMS can quantify those species which cannot be accurately measured using FT-IR alone.

2.1.1.2 Facility Utilities and Equipment Design

As mentioned earlier, equipment-level data alone does not provide a complete picture of the energy use in wafer fabrication. When equipment data is used to compose the LCI, the energy and water use associated with the utilities delivered to the manufacturing tool must also be accounted for. These facility utilities include ultrapure water (UPW), process cooling water (PCW), nitrogen, clean dry air (CDA), and the cleanroom environment delivered by the heating ventilation and air conditioning (HVAC) system. The industry standard SEMI S23 which was developed primarily for equipment life-cycle costing may be used to estimate the power demand and water use associated with these facilities.

SEMI S23 is a "Guide for Conservation of Energy, Utilities and Materials used by Semiconductor Manufacturing Equipment" created by the semiconductor industry association Sematech and defines a set of energy conversion factors with the general form:

$$\text{ECF}_{utility} = \text{Electrical energy (kWh)}/\text{Unit of Utility}(\text{m}^3 \text{ or kWh})$$

S23 presents energy conversion factors ($\text{ECF}_{utility}$) for electricity, UPW, PCW, facility nitrogen, CDA, house vacuum, exhaust, and HVAC as a function of the heat load of the other utilities.

2.1.1.3 Emissions Factors for Electricity Generation

Values for the greenhouse gas emissions associated with electricity can vary from nearly zero in certain places, such as Brazil (blessed with significant hydropower resources) to above 1 kg CO_2eq/kWh in the most coal-reliant geographies such as China and parts of Eastern Europe. Within countries and regions producing ICs, the values range from Chinese production (near 1 kg CO_2eq/kWh) down to lows of Costa Rica (about 0.08 kg CO_2eq/kWh). Given a potential order of magnitude difference in the carbon intensity of electricity, and because a large proportion of impacts occurring in wafer production are due to emissions from electricity generation, impact totals are very sensitive to the electricity emissions factor used in the LCI. For these reasons, particular care should be taken to use a representative and accurate emissions factor for electricity in LCI of wafer fabrication.

2.1.2 Government and Policy Methods Applicable to Semiconductor LCI

PFCs are an important group of emissions from semiconductor manufacturing due to their high infrared absorption, long lifetimes and consequential global impact. These compounds are used in wafer etching and post-deposition chamber clean and include CF_4, C_2F_6, NF_3, and SF_6, among others. In addition, a variety of fluorinated ethers, alkanes, amines and aminoethers are used as heat transfer fluids [39, 124] in equipment cooling, and N_2O is used in certain deposition reactions. A table of the global warming potential (GWP) impact factors for some of the PFCs used in wafer fabrication are shown in Table 2.1 [63]. These chemicals have impact factors many thousands of times that of CO_2. As such, global warming impacts are an important impact category to consider in the production of ICs, and, as noted in Chap. 1, previous LCA of semiconductors indicate the direct emissions of PFCs from wafer fabrication are an important contributor to this impact category.

Table 2.1 Global Warming
Potential (GWP) values
for some PFCs used
in semiconductor
manufacturing [63]

Chemical formula	GWP
NF_3	17,200
CF_4	7,390
CHF_3	14,800
C_2F_6	12,200
C_3F_8	8,830
CH_4	25
SF_6	22,800
N_2O	298
C_4F_6	$0.1{-}2^i$

Note: GWP values reflect 100
year time horizon^i undefined under
IPCC AR4

Because the climate change potential of PFCs is a global issue, the UN
has developed models for predicting or estimating the GWP of emissions from
semiconductor wafer fabrication operations, for use in regulation and voluntary
policy.

The United Nations Intergovernmental Panel on Climate Change (IPCC) has
developed tiered methods which may be selected on the basis of data availability.
The methods range from the least rigorous estimation scheme, requiring minimum
data (Tier 1), to the most rigorous and data-intensive method (Tier 2a). Here we
will review the Tier 2a method as described the 2001 IPCC Good Practice Guidance
document, which requires process-specific parameters.

IPCC Tier 2a Method - Process-specific parameters
Emissionsof FC_i ($E_{i,p}$):

$$E_{i,p} = (1 - h) \sum_p \left(\left[FC_{i,p} \left(1 - C_{i,p}\right) \left(1 - a_{i,p}\, d_{i,p}\right) \right] \right) \tag{2.1}$$

Where:

p = Process or process type (etching or CVD chamber cleaning).
$FC_{i,p}$ = kg of gas i fed into process/process type p (CF_4, C_2F_6, CHF_3, C_3F_8,
c-C_4F_8, NF_3, or SF_6).
h = Fraction of gas remaining in shipping container (heel) after use.
$C_{i,p}$ = Use rate (fraction destroyed or transformed) for each gas i and pro-
cess/process type p (in kg).
$a_{i,p}$ = Fraction of gas volume fed into in processes with emission control
technologies (company- or plant specific).
$d_{i,p}$ = Fraction of gas i destroyed by the emission control technology. If more
than one emission control technology is used for process/process type p, $d_{i,p}$ is
the mass-weighted average of the fractions destroyed by those emission control
technologies.

Table 2.2 IPCC Tier 2c
standard values [62]

	h_i	$1 - C_i$	B_i	d_i
CF_4	0.1	0.8	NA	0.9
C_2F_6	0.1	0.7	0.1	0.9
CHF_3	0.1	0.3	NA	0.9
C_3F_8	0.1	0.4	0.2	0.9
$c\text{-}C_4F_8$	0.1	0.3	NA	0.9
NF_3	0.1	0.2	NA	0.9
SF_6	0.1	0.5	NA	0.9

By-product emissions of CF_4 for $FC_{i,p}$ ($BE_{i,p}$):

$$BE_{i,p} = (1 - h) \sum_p \left[B_{i,p} \ FC_{i,p} \left(1 - a_{i,p} \ d_{CF4,p} \right) \right] \qquad (2.2)$$

Where:

$B_{i,p}$ = Fraction of gas i transformed into CF_4 for each process type p.
$d_{CF4,p}$ = Fraction of byproduct CF_4 destroyed by the emission control technology used for process type p.

Values used in this method are listed in Table 2.2.

This method provides a way to estimate emissions from a manufacturing facility without performing emissions measurements on site, and using only facility data for chemical purchasing. The definition of this methodology also allows such estimates to be standardized and comparable.

The US EPA developed a technique for estimating the annual mass of pollutant emissions at the scale of the US economy [17]. The EPA method uses data on facility production capacity, combined with factors for PFC emissions per unit of silicon area (area of product) specific to the number of layers of active area in the device being manufactured. The mass-balance method used to determine the annual emission of fluorinated greenhouse gases in the EPA model follows the IPCC tier 2c method [62].

From the EPA:
PFC emissions in year y (PFC_i):

$$PFC(y) = \sum_{\tau(y)}^{T(y)} \langle e_{\tau(y)} \rangle \ S_{\tau(y)} \qquad (2.3)$$

PFC (y) is the total for US PFC emissions in year y, for all PFCs from semiconductor processes:

$\langle e_{\tau(y)} \rangle$ is the population average emissions factor for linewidth technology τ.
$S_{\tau(y)}$ is the silicon consumed in producing linewidth technology τ in year y.

This method allows coarse modeling of a very large set of manufacturers, using a small set of well-designed, technology-specific emissions factors, combined with economic data which is available through public reporting (silicon purchasing data).

2.1.3 Adaptation of Existing Semiconductor LCI Data

While it is possible to develop an LCI for wafer fabrication using original data collection and estimation, LCA practitioners can also estimate the impacts associated with the production of a specific IC from published data in the literature. In this section, we will show how to tailor published LCI data for wafer production to estimate impacts related to a specific device. Care should be taken to match the subject of analysis with the functional unit of the previous study. When consulting the literature, be sure that the subject of the study found matches your device of interest in the following ways:

Product type: Is the integrated circuit a logic or memory device? If the device is memory, what type is it (DRAM, flash, EEPROM, etc.)? If the device is logic, is the product an analog or digital circuit? Computer processors and microcontrollers are normally digital circuits. If the product is a power or communications circuit, it may be analog.

Transistor configuration: Is the underlying transistor design a CMOS (complementary metal, oxide, semiconductor), bipolar or another configuration? Most digital ICs are currently built on CMOS-based circuits. Until recently, most analog circuits were built on bipolar transistors, which have been in some cases replaced by a CMOS-like bipolar implementation referred to as Bi-CMOS.

Year of production or technology node: A "technology node" is a term used to commonly refer to a set of production technologies specific to a device type (CMOS logic, DRAM and flash memory) and production year. The term was originally defined within the International Technology Roadmap for Semiconductors (ITRS) to allow a common shorthand within industry. The definition of the term "technology node" has changed over time, but the relevant definition for your year of production can be found in the ITRS report or update for that year [104]. If an older technology LCI is scaled up to represent a later IC, the results would likely overestimate in many impact categories if it is scaled per number of transistors, and may underestimate if scaled by area.

If the technology node is not known, then among chips of the same product type, year of production is an acceptable proxy.

Die size: The area of bare silicon die can be found through product specifications, or by physically grinding down the surface of a packaged chip to reveal the silicon device. The die surface is sometimes referred to as the "active area" of the chip. The die size is distinct from the package size, and may occupy anywhere from less than 10% to more than 80% of the package area.

Chip size: The dimensions of the packaged chip may be measured or determined from the chip's specifications.

If your functional unit matches the above parameters of the study, that study's results will be a good fit for your LCI. If the die size or chip size do not match, then it may be possible to adapt the study by normalizing its results per die area and chip area, if the front and back end LCI are reported separately in the study. Normalizing only on the basis on the external chip dimensions is dangerous, as there is wide variation in the fraction of chip size occupied by the die. If a previous study of an IC with a high die-to-chip area ratio is normalized per chip area, it will overestimate results for chips with lower die-to-chip area ratios.

2.1.4   Use of Economic Input-Output LCA for Wafer Fabrication

Economic Input-Output LCA (EIO-LCA) modeling is a useful tool for estimating impacts on a per dollar basis for particular industrial sectors [54]. EIO-LCA is particularly useful in quantifying impacts for large, stable commodity industries, in which products of the sector category are uniform, interchangeable or may otherwise be represented well by average pricing and environmental impact data across the sector. Because the energy consumption is easily tracked using economic data, EIO-LCA is most effective for sectors in which environmental impacts are predominantly energy-related.

Because the semiconductor industry represents a wide variety of products, from light emitting diodes to high-end computer processors, the aggregate results available through EIO-LCA are not the most accurate for most products. Also, because of the rapid technological change in the industry, product types, prices, energy use and emissions per unit economic value may not be expected to be stable from year to year. For both of the above reasons, EIO-LCA is not an ideal tool for estimating impacts from semiconductor fabrication. In addition, unless an EIO-LCA model is specifically modified to include direct emissions from semiconductor manufacturing, EIO-LCA does not provide complete LCI data for semiconductor manufacturing.

Economic data and EIO models are very useful for cross-checking specific results from other methods of LCI development, such as average values for silicon consumption or energy consumption per unit of economic value. These comparisons can also be made without the step of economic normalization, allowing the uncertainties associated with product pricing to be eliminated from the comparison. For example, using census data on the sector's annual consumption of silicon wafer area and total annual energy consumption, an average value for energy use per unit area of silicon can be determined for comparison [2].

2.2   LCI of Chemical Production

In one early, influential paper on LCI of semiconductors, Williams presented an LCI of a memory chip and hypothesized that energy consumption in production of the high purity chemicals used in wafer fabrication would contribute considerably to the life-cycle environmental impact of ICs [130]. Other studies have attempted to address this hypothesis, and one conference paper in particular has shown chemicals production to be a substantial fraction of energy use in production [55x], but as there is still a lack of publicly-available LCI of the high purity gases used in wafer processing, this conjecture cannot be repeated or substantiated for additional cases. Based on the available evidence, it seems that the production of chemicals is large enough that it should be accounted for as a part of the life-cycle of an IC.

In Williams' analysis of a memory chip, the author stated that chemical production and purification likely contributes a considerable amount to life-cycle energy consumption, yet due to a lack of representative data, the question of how much could not be definitively resolved [130]. In that study, the LCI data used for process chemicals production (typically in the purities of 99.999% to 99.9999995%) was based on LCI data representative of industrial grade (90–99% pure), and was therefore considered a lower bound. Nevertheless, even using these low estimates, the LCI for production of chemicals amounted to over 4% of total life-cycle energy consumption. It is important to keep in mind that the subject of that study, a DRAM chip, had low use phase power consumption and thus a lower use phase contribution to life-cycle impacts, causing production impacts to be of higher relative importance. For ICs with higher use phase power, for example logic ICs, the upstream impacts would have a lower percentage contribution.

In a later study from Intel, a closer investigation of the energy associated with chemicals production is made as a part of a paper concerning boundary definition for semiconductor carbon footprint [55x]. Due to the large number of different chemicals used in production, LCI data collection for all chemicals was not possible, and the study defined a mass threshold for inclusion in the LCI, limiting the total number of materials considered to only 30. Unfortunately, by definition, the low-volume materials which are excluded are the most exotic, expensive and most highly processed chemicals used in manufacturing. Therefore, although the study accounts for 99% of the total mass of chemicals and gases used, the author states that there is uncertainty remaining about how much of the carbon impact is excluded when a mass threshold is used.

The author, Higgs, states that the LCI data for most chemicals included in the study represent standard, industrial grades rather than the actual high-purity grades used, but that for some gases, the existing LCI datasets were modified to reflect representative, high-purity electronics grade materials. It may be assumed that for some bulk gases purification is done on site, and that this aspect of the LCI for some gases could be accounted for as a part of facility data collection, which would be a high quality data source. One valuable finding from the paper is that, for those chemicals modeled, purification to electronics grade required a CO_2 impact of 20–1,000% higher than the initial industrial grade. While the effort made to account for purification of process chemicals was commendable, and provides valuable insight into the range of values potentially associated with data gaps, purification modeling was only possible for a few materials in this study, and so the baseline LCI totals for chemicals production may still underreport. The missing purification data is addressed with an upper bound, which is estimated using the modeled purification process data as a proxy for purification of the remaining high-purity chemicals. The report indicates that, using these LCI methods, the upstream production of chemicals contributes 23–28% of GWP impacts in semiconductor production.

Despite further inquiry on this subject, including some attempts to develop LCI data for high purity chemicals using theoretical models [74], the lack of publicly-available LCI data for the high-purity gases and specialized chemicals used in wafer fabrication limits the ability to model this life-cycle stage.

2.3   LCI of Infrastructure

Semiconductor fabs are large, extremely expensive capital projects. The cost of building construction and equipment for a new fab topped one billion dollars in as early as 1997 [46], and can currently exceed 10 billion dollars. It is clear that a significant amount of economic activity and material use is involved in the development of a wafer fab, and it is tempting to account for this aspect of semiconductor fabrication using LCI data from the construction industry, but fab construction is not representative of an average construction project. Possibly because aspects of the cost incurred in fab construction are intellectual, non-physical costs, if construction LCI data is applied per dollar of activity, it would be representative of a much larger physical facility. The potential to over-predict impacts is amplified when impacts are depreciated over the useful life time of a fab, which can by typically 10–15 years, rather than the lifetime of an average building, which can be 20–50 years. Indeed, the total potential lifetime of the fab in a secondary use as a non-wafer fab building may be the same as any other large building. This uncertainty about fab lifetime and the absence of representative LCI data for a semiconductor fab construction make quantification of this life-cycle stage difficult. In the LCI presented in this paper, economic input-output data for the construction industry is used as an upper bound for impacts in this life-cycle stage.

PAS 2050, the British standard which outlines the methodology required for LCA in environmental reporting, does not require accounting for capital infrastructure, such as the building and equipment used in production. For this reason, as well as the uncertainty in evaluating the indirect impacts of fab construction, this life-cycle stage is commonly omitted from analysis.

2.4   Modeling Energy Consumption in the Use Phase

Operational energy efficiency at the CPU has shown dramatic improvement over the past several years, yet brings continued challenges, as operational power is inextricably linked to IC performance. While power consumption per transistor falls with scaling, the total number of transistors per chip increases to deliver greater performance in each successive generation. Chip-level power management can be governed by a wide variety of choices made in the definition of an IC product's process technology, beginning with the choice of substrate (silicon-on-insulator vs. bulk silicon wafers), through to the choice of packaging and design for heat dissipation. As the process technology choices which affect chip power consumption also affect performance, and because these process technologies are continuously evolving, it is important to take care in assumptions made for power consumption in the use phase of an LCI model. A high-performance logic chip from one year cannot be expected to have the same power ratings as an equivalent product just two years earlier.

In many cases, the use phase is the dominant source of environmental impacts for integrated circuits among the life-cycle stages. Some ICs, like sensors and some types of memory, operate at very low power or are active with rare intermittency. For these products, use phase energy consumption may be modest and the manufacturing phase becomes the primary concern. Nevertheless, apart from of these exceptions, the use phase is an important life-cycle stage for the broad majority of ICs for use on most electrical grids throughout the world. Most integrated circuits which consume more than a few watts in operation will have a use phase which dominates life-cycle energy consumption.

The relative impacts of the use stage are dependent on not only the power consumption of the chip, but the emissions factor for electricity at the location of use as well as other factors which come into play at the level of the electronic product (power supply efficiency, availability of product idle and standby states, and software-integrated power management) and the use phase scenario (the frequency and intensity of use).

2.4.1 Specifying the Use Phase Scenario: Location, Time in Operation and Utilization Rate

The use phase scenario will define the location, frequency and length of use, as well as the utilization rate being demanded from the IC.

Depending on the electricity mix in the location of use, when use phase operation is the largest consumer of primary energy resources, it may or may not also be the largest contributor to GWP, acidification, ground-level ozone formation, water demand and other environmental impacts. When use occurs in a location with an exceptionally low emissions factors (e.g. Brazil, 73 g CO_2eq/kWh) or water consumption factors for electricity, other use phase impacts besides primary energy consumption are mitigated. Because there is such a wide range of emissions factors for electricity among all of the electrical grids in the world, the range of potential impacts for a certain chip's operational life can span over many orders of magnitude depending on the location of use. The choice of a representative emissions factor for electricity in the use phase, and understanding of the location or mix of use locations, is thus very important to an accurate life-cycle assessment for IC.

Frequency and duration of use are intuitive to grasp and are commonly addressed in LCA of any energy-using product. The total time in operation will be defined by the number of hours of active use per day, number of days per year, and years in the product's lifetime. Warranty reports, when available, can be a good source of information on the average lifetime of a product group, such as laptops or netbooks. In addition to the time in active operation, an IC in most cases will spend time in other operational modes, such as stand-by or idle. Because, in many cases, an IC will spend more time in idle or standby than in active operation, it is important to also quantify the energy consumption (the period of time and power consumption) in these other modes.

The utilization rate, the average fraction of the maximum power used in a given application, is a less intuitive but equally important aspect of an IC's use phase. Unlike most other electric products, the rated power for a logic chip will often be the maximum power demand. In normal operation, the chip will almost never require the maximum rated power. Instead, an IC will use the necessary fraction of its total computational, memory, control, or communication resources, and demand the associated fraction of its rated power. On average, for example, a central processing unit (CPU) may only use 15% or 30% of its rated power.

The utilization rate or application rate of the chip will be defined by the use case. For memory, the power required by the chip to deliver streaming video is different from the power used to support the memory demands of a typical office application suite. Similarly, a processor operating in a datacenter server will employ a much higher fraction of its computational power on average than one in a laptop. There are a different energy efficiency or power consumption benchmarking tests for different IC device types, and these testing procedures can change frequently to adapt to new circumstances (i.e., changes in software, communication standards or peripheral devices). One example for central processors and chipsets is the SYSMark standard.

2.4.2 System-Level Factors

Because most any chip which is being used in a computer, laptop or server is being delivered DC power through a power supply, the efficiency of that power supply's conversion from AC to DC power has a large influence over the chip's use phase energy consumption. If, for example, a power supply has an efficiency of 74%, the chip's use phase energy demand, when considered in the context of the integrated product, will be 1.35 times the chip's direct power demand. In a computer with a 90% efficient power supply, the same chip would demand 1.11 times the chip's own energy demand, or 18% less energy over its life-cycle.

If the functional unit of study is used in a server, other system-level factors which arise at the level of a data center may be of concern. For example, the IC's excess heat dissipation will affect the heat load in the data center, and therefore the center's cooling requirements and associated energy consumption.

2.4.3 Software

In the early 2000s, operating system-integrated chip power management (aka, advanced or dynamic power management) was introduced, which allowed software to hibernate or shut down the CPU when the user was inactive. Windows 2000 and XP both integrated advanced power management but neither of these versions of Windows had advanced power management settings enabled by default, and the functionality was often not enabled by the user. In a 2007 market research study, as many as 60% of computer users in the US did not shut down their computer at the end of the day, resulting in the needless emission of roughly 14 million tons of CO_2 that year [5]. In Windows 7, the default settings for shipment were for lower power consumption, which supported wider use. The default setting, rather than the available functionality, is a critical aspect of power management, and also of the life-cycle impacts of an electronic product.

The influence of software over the life-cycle impacts of computing also extends into other life-cycle stages beyond the use phase. Software's demand for increasing memory and computational power drives the obsolescence of a computer, specifically through the demand for more advanced logic and higher capacities of DRAM. By driving ICs into retirement, ever bulkier and more memory-intensive operating systems and applications not only trigger the impacts associated with EoL to occur at an earlier point in time, but also obligate production of their replacements. The role of software in the lifetime of electronics is easy to let slip into the background. In the case of LCA of ICs, software is usually not considered when the functional unit and boundary of a LCA of ICs is set, but given the influence of software over life-cycle impacts, this practice should be reconsidered where possible.

2.5 End of Life

The most evident end-of-life (EOL) impacts from integrated circuit chips are lead emissions, when the solder or wire-bonding materials contained inside the packaged chip or used to attach the chip to a board contain lead, a potent neurotoxin. EOL lead emissions have been mitigated since 2006, when the EU's Restriction on Hazardous Substances (RoHS) regulation banned the sale of products containing leaded solders in the EU. Studies of EOL electronics have identified and quantified other end effects of computer disposal through modeling as well as measurement of toxic compounds present near informal recycling sites [21, 28, 132, 133]. These other toxins (dioxins, brominated flame retardants, and other bio-accumulative or toxic materials) largely represent emissions from the breakdown or combustion of a computer's more massive components, such as the printed wiring board and plastic housings. While there may be harmful emissions besides lead from the decomposition of a logic chip, these have not yet been identified or measured.

Chapter 3
Semiconductor Manufacturing Trends in Product Type and Geography

3.1 Introduction

This chapter provides an overview of the semiconductor industry, in order to establish the dominant device types, and to identify environmentally relevant trends in semiconductor product types and geographic concentration of manufacturing capacity.

The complementary metal oxide semiconductor (CMOS) configuration is the prevailing transistor structure in production today. CMOS is the standard structure for digital logic and CMOS transistors, integrated in various forms with capacitors, form DRAM and flash memory. Although an exact accounting of the share of worldwide semiconductor capacity dedicated to CMOS logic, in particular, is difficult to ascertain, by extrapolating from market data concerning product types, it seems that more than half of the semiconductor market is CMOS logic. CMOS logic is therefore the primary subject of analysis in this thesis.

There is a significant expansion in semiconductor capacity in Singapore, Taiwan, Malaysia and China, which can be expected to result in an increased carbon intensity of ICs. Neither Singapore, Malaysia nor China are signatories to the Kyoto Protocol and while industry consortia in Japan, Europe and the US have announced GHG emissions reduction goals, industry in China, Taiwan, Singapore and Malaysia have been silent on the issue. Based on anecdotal evidence, the abatement of per-fluoro-compounds (PFCs, potent global warming gases used in production) is not common in Singapore or China. As production capacity shifts from Europe and the US to non-Kyoto-bound East Asian countries, more and more production could be conducted without controls on PFC emissions. Relocation of wafer fabrication capacity to China in particular has the potential to drastically increase the environmental impact profile of production as a result of the high global warming intensity of electricity in China. Finally, while often overlooked in LCA, transportation plays an important role in life-cycle impacts for semiconductors, as

S.B. Boyd, *Life-Cycle Assessment of Semiconductors*, DOI 10.1007/978-1-4419-9988-7_3,   27
© Springer Science+Business Media, LLC 2012

will be illustrated in Chap. 3. The lengthy transportation necessary to take products from the fabrication site to the US and Europe (the largest end markets) also worsens the profile of ICs produced in East Asia.

The growing popularity of flash memory is significant from an environmental perspective due to the unique materials used in its production and their consequential emissions. The use of flash in mp3 players, memory sticks and other portable storage has expanded the flash market dramatically over the last several years. As the storage density of flash memory increases and its cost falls, flash has become competitive with hard disk storage. In the coming years, as solid-state storage replaces hard disks in portable computers, the flash market can be expected to grow further. For this reason, the environmental analysis of flash memory is also chosen as a topic for this thesis.

3.2 The Semiconductor Industry: Size, Growth and Trends

In 2008, worldwide semiconductor industry revenue amounted to $260 billion, with production averaging 1.9 million wafer starts per week in terms of 8 inch wafer equivalents [112, 115]. This amounts to over 5 billion square inches, or 800 acres, of silicon area produced in that year.

Growth in the global semiconductor market has occurred with vitality and volatility throughout the history of the industry, with year-over-year revenue growing or contracting at rates between −32% and 42% since 1993 [113]. The industry has generally sustained a high rate of growth overall, with a compound annual growth rate (CAGR) of 8% over the 1998–2007 period (in then-current US dollars). This represents a slowing from 12% for 1998–2004, and 17.5% for 1962–1995. The variable but strong expansion of the industry is illustrated by global annual revenues from 1952 to 2007 in Fig. 3.1 [113].

3.2.1 Technology Scaling

Throughout the history of the industry, the processes used to fabricate semiconductor circuits have continuously changed to meet scaling and performance goals. The common metric used to describe the scale of a given circuit generation, the "technology node," is equivalent to the DRAM half-pitch, which is based on the average line width and space between lines connecting DRAM cells. This metric is based on the lithographic process because, historically, the advance from one technology node to the next was determined by lithographic technology. For example, the introduction of deep-UV in the late 1980s enabled the development of 0.5 micron node DRAM [8]. More recently scaling has been constrained by the ramifications of reduced feature size on the performance of the transistor, and

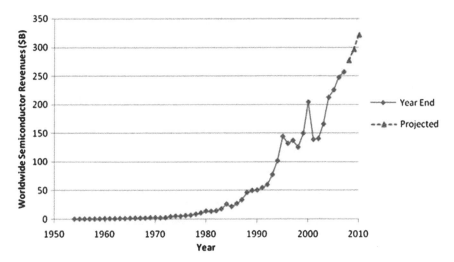

Fig. 3.1 Global annual revenues from 1952 to 2007

the ability of process engineers to overcome these effects. An early example of scaling-related performance limitation is the parasitic capacitance and cross-talk in metal lines which occurred due to the insufficient insulation between neighboring metal lines. This issue was overcome with the use of advanced low-permittivity (low-κ) dielectrics in the lower, denser interconnect lines. More recently, sub-threshold leakage current, increased junction leakage, gate tunneling and other deleterious effects have necessitated further material and process innovations, such as substrate biasing, source drain extensions, raised source/drain geometries, high-κ gate dielectrics and many other process innovations. These ongoing design and process improvements have roughly doubled the transistor density of low-cost circuits every 24 months, allowing Moore's Law to hold, for over 35 years.

As each new technology generation is released, additional production capacity is either built or converted from older manufacturing facilities. This ongoing conversion of manufacturing capacity is illustrated by the volumes of capacity per each technology node, for the years 2005–2007, shown in Fig. 3.2.

3.2.2   Dominant Circuit Elements: The Transition from Bipolar to CMOS Transistors

The first transistor, designed at Bell Labs in 1947, was a bipolar junction transistor (BJT). Bipolar phased out of digital logic in the late 1970s, replaced by faster n-type metal oxide semiconductor field effect transistors (n-type MOSFET or NMOS). In the mid-1980s, the complementary MOS (CMOS) structure became

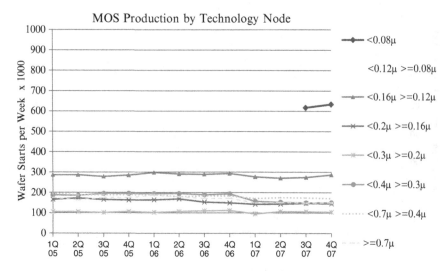

Fig. 3.2 Worldwide MOS production capacity by technology node

the dominant element of digital logic due to the advantageous features of lower static power consumption and higher allowable circuit density inherent in the CMOS structure.

Bipolar comprised only about 7% of semiconductor device production capacity in 2007 [114], with the remainder dedicated to MOS-based circuits. The small fraction of fabrication capacity is a reflection of the dominance of MOS in digital applications as well as the replacement of bipolar transistors in some traditionally bipolar applications. Bipolar elements have a high output resistance and are thus particularly useful for amplification and switching, making them the typical the building block of analog IC and power components. More recently some traditionally bipolar-based analog applications such as radio frequency (RF) communication have been implemented completely in CMOS, however. For example, double-diffused MOS (DMOS), a power transistor design based on the CMOS structure, are used in power supplies and low-voltage motors. In mixed-signal devices (e.g., cell phone chips), bipolar transistors are now frequently used in conjunction with CMOS transistors using a combined process flow known as bi-CMOS. Though their higher allowable voltage make bipolar transistors critical for certain power and signal applications, bipolar production of has been declining consistently for several years, as shown in Fig. 3.2.

CMOS transistors are the most common form of MOS and the dominant basic circuit element in random access memory (RAM), microprocessor units (MPU) and other digital logic chips. Nearly all digital logic and most current dynamic random access memory (DRAM) and electrically-erasable read-only memory (EEPROM, a.k.a. flash) technologies are also based on CMOS transistors.

3.2.3 Products

The semiconductor market is split among several major product types, with the largest fractions being MOS logic (composing 26% of the market), MOS micro-processing units (MPU) (14% of the market), analog circuits (14%) and MOS micro-computers (MCU) (9%). DRAM and flash EEPROM memory comprise 13% and 9% of the market respectively, though the flash EEPROM market is expected to grow rapidly in the next few years, and the DRAM market is contracting.

The fastest growing market segments are flash memory, analog IC and MPU. The projected growth for flash revenue during the 2007–2010 period is 20.0%, and in 2010, it is forecast that flash will contribute to 13% of the global market. The growth of flash memory has been driven by the increasing use of the technology in consumer electronics such as MP3 players. Analog IC and MPU are projected to grow by 8.3% and 8.1% respectively during the same period. The MPU market follows computer sales, which are expected to remain strong in the foreseeable future. Analog ICs are used, along with other logic components, in phones and other mobile communications devices. The analog IC market sector has thus been one of the fastest growing product sectors in recent years, due to increased demand from mobile consumer electronics manufacturers [3]. As more and more people use mobile devices to perform many of the same tasks traditionally carried out on a personal computer, the analog IC sector may be expected to have strong growth in the longer term.

In 2010, the DRAM market is forecast to contribute to 11% of global revenues, down from 13% in 2007 [113]. While revenue for this sector is not expected to decline in 2007–2010, the forecast growth rate is only 1.5%, the lowest of any of the product sectors. Flash memory has replaced DRAM in certain consumer electronics products, contributing to this slowing in growth. Also, the price of DRAM has experienced volatility and extreme price drops in recent years due to spates of overproduction. DRAM is a component necessary for personal computers and due to the relatively stable demand from the computer market, the DRAM sector is not expected to contract in the next several years.

Other classes of semiconductor products include discrete components (e.g. individual transistors, rectifiers, diodes and sensors), optical components (e.g. LEDs, charge coupled devices, light sensors, laser devices, character displays) and analog electronics (e.g. power supply components, op-amps, automotive control chips, radio frequency (RF) communication chips, telecommunications equipment components). The composition of global semiconductor market revenue in 2007 is illustrated in Fig. 3.3.

In order to clarify the relative popularity of the various device types, the Table 3.1 lists product categories and their associated device types by their percentage contribution to world revenue. The product types of MOS logic, MPU, and MOS micro, shown in Fig. 3.3, are composed most commonly of CMOS logic. CMOS logic, as a fundamental circuit type, composed the largest fraction of world semiconductor production in 2007.

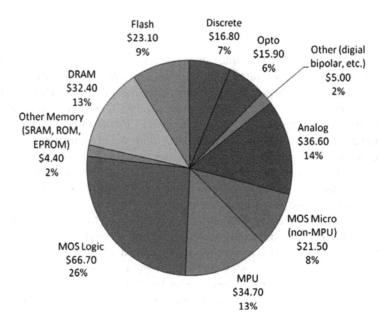

Fig. 3.3 2007 Worldwide revenues by product type

Table 3.1 Semiconductor products and their dominant device types, by market volume

Product category	Percentage of world revenue	Dominant circuit element
MOS Logic	26%	Primarily CMOS, also NMOS, PMOS or BiCMOS
Memory	23%	CMOS, NMOS, PMOS or BiCMOS
Analog	14%	Bipolar, MOS or BiMOS
MPU	13%	CMOS
MOS Micro (non-MPU)	8%	Primarily CMOS, MOS or BiMOS
Discrete	7%	Diodes, MOS, bipolar
Opto	6%	Diodes, MOS, bipolar
Other IC	2%	Bipolar, MOS

3.2.4 Geographic Concentration of Production

The semiconductor industry extends globally and many companies operate internationally. The dominant model of operation among US-based manufacturers is to locate research and design activities in the US, wafer manufacturing at another site, and chip and component assembly in a location with a low cost of labor. The location of wafer manufacturing is usually not based on labor costs, but rather a combination of factors, including tax incentives, availability and cost of capital, access to reliable power and water supplies and the ease of the regulatory environment [16].

During the last decade, some semiconductor companies have chosen to outsource some or all of their wafer production to foundries rather than owning their own fabrication facilities. This trend in outsourcing has been driven by the increasing capital cost of wafer fabrication plants ("fabs"). The increasing complexity of process flows requires that each fab contain more individual pieces of equipment, while increases in wafer size have increased equipment footprints and thereby the overall size of each fab. A typical 200mm CMOS logic wafer fab in 1997 would have cost about $800 million, while an advanced 300 mm wafer fab in 2001 cost approximately $2 billion and a fab in 2007 could cost as much as $18 billion [64]. Approximately 14% of worldwide manufacturing capacity is held by foundries with a concentration in Southeast Asia [114]. The largest contract foundries are Taiwan Semiconductor Manufacturing Corporation (TSMC) and United Microelectronics Corporation (UMC), which are based in Taiwan, Chartered Semiconductor of Singapore and Semiconductor Manufacturing International Corporation (SMIC) of China. Some major semiconductor and consumer product companies are outsourcing all of their production to foundries. Examples of this strategy are the "fabless" Qualcomm, Broadcom, LSI Logic and Nvidia. Other companies only use foundries selectively, for older products with well-established production processes and lower margins, or for products which are not related to the technical competence of the firm. STMicroelectronics, Analog Devices and Freescale each have about 20% of their production performed externally.

SIA worldwide revenue reports show dramatic growth in East Asia (Fig. 3.4). Although revenue and sales information related to a company or market can give a broad overview of trends in the industry, it cannot be used as a proxy for information concerning the location and volume of actual wafer fabrication capacity. For example, Intel is headquartered in California but owns fabs in 12 countries around the world, and, although TSMC is based in Taiwan, the company runs foundries in Singapore, Taiwan, China and the US. Also, revenue figures may not be representative of actual production, as the price of certain semiconductor products, such as DRAM, can experience extreme drops in price, often due to overproduction. Industry reports describing capacity, however, also cannot describe the production volume of any given product type with certainty because many companies produce a range of similar products in the same facilities. For example, most companies who produce DRAM also produce flash memory in the same fab, and some contract foundries will produce a set of product types, such as CMOS logic and bi-CMOS, in the same fab. Thus reports concerning capacity cannot always be representative of the actual product being produced as these manufacturers switch between product types in order to maximize profit and capacity utilization. Industry reports on revenues or capacity can give an overview of general trends, but, for the purposes of determining the production volume of a given product type, must be interpreted with an understanding of the production dynamics internal to each firm. With these qualifications in mind, the industry trend of fabrication relocation to, and capacity growth in, Taiwan and China is evident from company-specific research.

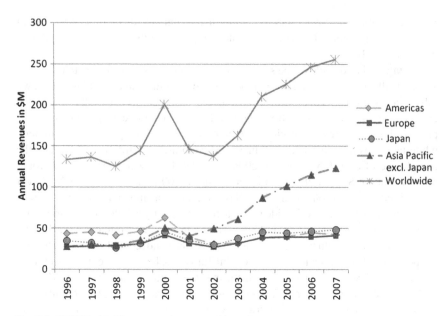

Fig. 3.4 2007 Worldwide revenues by geographic region

There is a clear expansion in wafer manufacturing in Taiwan and China which is occurring through the growth of Chinese and Taiwanese foundries, as well as the relocation of manufacturing capacity by established firms to China and joint development agreements between European or US firms and Taiwanese or Chinese firms. Chinese wafer fabrication companies which have plans to build new capacity in China between 2008 and 2011 include SMIC (Semiconductor Manufacturing International Corp.), He Jian Technology, Grace Semiconductor, Hua Hong NEC, Shanghai SIM-BCD, Jilin Sino-Microelectronics and CSMC (Central Semiconductor Manufacturing Corporation) Technologies Co. These firms operate for the most part as contract foundries, with some design and assembly services and represent capacity that is being out-sourced from companies largely in the US. Some established companies are also expanding their existing capacity or building fabs for the first time in China and Taiwan. A prominent example of the latter is Intel's plan to build a $2.5B, 90-nm node, 300 mm fab by 2010 in Dalian, on the coast of Northern China [22]. Other examples include Philips Electronics' partnership with Jilin Sino-Microelectronics Co. Ltd (JSMC) through a joint firm named Philips JiLin to produce bipolar power controllers, and the partnership between ST Microelectronics, a Swiss company, and Hynix, a Korean company, to build a fab in China. (Due to Chinese government regulation, the creation of a partnership or joint venture between a foreign and Chinese firm is much more common than an investment in wholly owned manufacturing facilities in China by a foreign company.)

The growth in capacity located in Taiwan and China is underscored by trends in semiconductor manufacturing equipment sales. The semiconductor equipment market grew by 20% and 14% in Taiwan and China, respectively [135]. This revenue data includes sales of chip assembly as well as wafer processing equipment, but wafer equipment purchases accounted for 81% of the Taiwanese semiconductor equipment market in 2007.

3.3 Conclusions

Semiconductor products are an important topic for life cycle assessment due to the magnitude and growth of the production of these products. CMOS-based digital logic, used in micro-processors and controllers, accounts for close to half of worldwide semiconductor production and represents the largest proportion of the market by device type. Due to its widespread production and use, CMOS logic is chosen as the first subject of investigation and represents the topic of study in the bulk of this thesis. Due to the rapid growth in production of flash memory, and the forecast that this product type will be the second largest product sector by 2010, flash memory is chosen as the second subject of study in this thesis. DRAM, which is forecast to be the third most widely produced product type in 2010, and is also a critical component to computers, is chosen as the third subject of study in this thesis.

The growth of new capacity and re-location of production to Taiwan, Singapore, Malaysia and China is also noted as an important topic of analysis, due to the lack of PFC-abatement policy in these countries as well as the high global warming intensity of electricity in China. In each LCA study in this thesis, the relative impacts of wafer production with and without PFC abatement are determined. In Chap. 3, which focuses on the energy-related impacts of CMOS logic, two geographical cases are presented, one in the US and the other in China, to clarify the influence of the electricity mix at the production site on total life-cycle global warming impacts.

Chapter 4
Life-Cycle Energy and Global Warming Emissions of CMOS Logic[1]

4.1 Introduction

Complementary metal oxide semiconductor (CMOS) is the dominant device structure for digital logic. The central processing unit (CPU) in desktops, laptops, handheld devices and servers, as well as nearly all embedded logic (the chips in appliances and toys) are composed of CMOS transistors. Every one to three years, a new generation or technology node of CMOS is introduced, based on design laws which have been established through industrial collaboration. Due to the cooperation necessary to plan and achieve the goals for each generation, there is considerable homogeneity among the devices manufactured by the major logic producers at each technology node. A generic version of CMOS may thus be used in LCA to represent logic products from many different manufacturers.

This chapter provides a life-cycle energy analysis for CMOS logic chips over 7 technology generations with the purpose of comparing energy demand and global warming potential (GWP) impacts of the life-cycle stages, examining trends in these impacts over time and evaluating their sensitivity to data uncertainty and changes in production metrics such as yield. Chips of generic CMOS logic, produced at a mature, high production capacity fab are evaluated at each technology node over a 18 year period, from the 350 nm node (circa 1995), to the 32 nm node (circa 2013). This study is composed of production-related life-cycle inventory (LCI) data, based on emissions measurements, process formulas and equipment electrical tests, combined with previously published LCI data for chemicals, electricity and water, as well as publicly available use-phase data for computer chips. A hybrid LCI model is used. Wafer production, electricity generation, water supply and certain materials are represented by process LCI data, while the remaining materials as well as the fab building and manufacturing equipment are described using economic input-output life cycle assessment (EIO-LCA) methods [18]. While life-cycle energy and GWP

---

[1] Portions of this chapter are reprinted with permission from S. Boyd et al. (S. B. Boyd, A. Horvath, D. Dornfeld), *Environmental Science & Technology* **2009** *43* (19), 7303-7309. Copyright 2009 American Chemical Society

of emissions have increased on the basis of a wafer or die as the functional unit, these impacts have been reducing per unit of computational power. Sensitivity analysis of the model shows that impacts have the highest relative sensitivity to wafer yield, line yield and die size and largest absolute sensitivity to the use-phase power demand of the chip.

4.2 Methodology

This LCA includes materials production, wafer processing, die packaging, transportation and use of the logic chip (Fig. 4.1). The LCA model is hybrid, using a combination of process-based LCA and economic input-output (EIO) LCA data (Table 4.1). The functional unit is one packaged die, but in order to allow further analysis and to investigate trends, results are also presented per wafer and per million transistors.

At end-of-life, it is assumed that there is no recoverable energy value in the chip. Other end-of-life GWP impacts or offsets can occur when the system of the entire computer or electronic product in considered, but are not included in this account because the functional unit of this LCA is the chip alone. Because there is no positive energy value and no known global warming impact of an IC at end-of-life, the net global warming impact in this life-cycle stage is zero.

4.2.1 Inventory Model Structure

In order to clarify the model structure and in order to demonstrate the sensitivity of results to variation in model parameters, the inventory model is described algebraically. The contributors to the life-cycle energy requirements (e_{total}) and global warming potential (GWP) of life-cycle emissions (g_{total}) are illustrated in (4.1) and (4.2).

$$e_{total} = e_{up} + e_{inf} + e_{prod} + e_{trans} + e_{use} \qquad (4.1)$$

Fig. 4.1 Life-cycle stages

Table 4.1 Summary of data sources		
Silicon	Process LCA
Chemicals	Process and EIO-LCA
Infrastructure and equipment	EIO-LCA
Fabrication	Process LCA
Electricity	Process and EIO-LCA
Water	Process and EIO-LCA
Transportation	Process LCA
Use	Process LCA

e_{up} : energy for upstream materials
e_{inf} : energy for infrastructure
e_{prod} : energy for production
e_{trans} : energy for transportation
e_{use} : use phase energy

$$g_{total} = g_{up} + g_{inf} + g_{prod} + g_{trans} + g_{use} \qquad (4.2)$$

g_{up} : GWP of emissions due to upstream materials
g_{inf} : GWP of emissions due to infrastructure
g_{prod} : GWP of emissions due to production
g_{trans} : GWP of emissions due to transportation
g_{use} : GWP of emissions due to use phase energy

Energy and global warming potentials for chemical production, i.e., upstream activity, are given by e_{up} in (4.3) and g_{up} in (4.5). Process chemicals are split into two sets, the first set of m chemicals for which EIO-LCA data are used [18], and a second set of q chemicals for which process LCA data are used. The term $e_j^\$$ is the energy consumption per dollar value of chemical j, c_j is the cost per unit mass of j and m_{ij} is the mass of j consumed in process step i. The wafer yield (Y_{wafer}) is the percentage of good die per wafer, the line yield (Y_{line}) is the percentage of finished wafers of those started, and gross yield (n_{die}) is the number of dice per wafer.

$$e_{up} = \frac{1}{Y_{wafer} Y_{line}\ n_{die}} \left[\sum_{j=1}^{m} e_j^\$ c_j \sum_{i=1}^{n} m_{ij} + \sum_{k=1}^{q} e_k^m \sum_{i=1}^{n} m_{ik} \right] \qquad (4.3)$$

$$(4.4)$$

$e_j^\$$: energy consumption per dollar value of chemical j
c_j : cost per unit mass of j
m_{ij} : mass of chemical j (from set of m chemicals having EIO data) consumed in process step i
e_k^m : energy consumption per unit mass of chemical k
m_{ik} : mass of chemical k (from set of q chemicals having process data) consumed in process step i

$$g_{up} = \frac{1}{Y_{wafer} Y_{line}\ n_{die}} \left[\sum_{j=1}^{m} g_j^\$ c_j \sum_{i=1}^{n} m_{ij} + \sum_{k=1}^{q} g_k^m \sum_{i=1}^{n} m_{ij} \right] \qquad (4.5)$$

$g_j^\$$: GWP of emissions per dollar value of chemical j
c_j : cost per unit mass of j
m_{ij} : mass of j consumed in process step i
g_k^m : GWP of emissions per unit mass of chemical k
m_{ik} : mass of k consumed in process step i

The energy and greenhouse gas (GHG) emissions due to facility infrastructure and capital equipment are found using the EIO-LCA method.CMU EIO-LCA method. This method is summarized by (4.6), where A is the $n \times n$ economy-wide transactions matrix, R is the $n \times m$ impacts matrix where r_{ij} is the ith type impact of sector j. The demand vector $\hat{c}_{inf}$ is an n item vector with the costs of construction and equipment at the positions corresponding to their industrial sectors in A. The resulting vector $\hat{b}$ is the set of life-cycle energy demands and global warming emissions of the sectors used in this study. The EIO-LCA model is described completely in the literature [53,54].

$$\hat{b} = R (I - A)^{-1} \hat{c}_{inf} \qquad (4.6)$$

The total energy used in production (e_{prod}) consists of all electricity and natural gas used by production tools and the facility infrastructure, as given in (4.8). Equation (4.9) describes the GWP of production which includes both energy-related emissions and direct GHG releases.

$$e_{prod} = \frac{1}{Y_{wafer} Y_{line} n_{die}} \left[\sum_{i=1}^{n} \left(\frac{t_i \, p_i + e_{CH_4} t_i \, m_{i,CH_4} + \hat{e}_w^v \, \hat{w}_i \, t_i}{\varepsilon_i \, u} \right) \right. \qquad (4.7)$$

$$\left. + \left(\frac{p_{hvac} + p_{lt} + p_{trt} + p_{cda} + p_{exh}}{c} \right) \right] \qquad (4.8)$$

t_i: duration of process i
p_i: process tool and POU abatement power consumption for process i, per chamber
e_{CH_4}: energy per unit mass methane
m_{i,CH_4}: mass flow of methane for process i
$\hat{e}_w$: vector of volumetric energy requirements for ultrapure, process cooling and city water flow, per chamber
$\hat{w}_i$: vector of ultrapure, process cooling and city water flows for process i, per chamber
ε_i: process utilization for process step i
u : tool utilization
p_{hvac} : facility HVAC power consumption
p_{lt} : facility lighting power consumption
p_{trt} : facility treatment system power consumption
p_{cda} : facility CDA power consumption
p_{exh} : facility exhaust system power consumption

$$g_{prod} = \frac{1}{Y_{wafer} Y_{line} n_{die}} \left(\sum_{i=1}^{n} t_i \hat{m}_i \hat{g} + l \right) + e_{prod} \, g_e^{prod} \qquad (4.9)$$

t_i: duration of process i
$\hat{m}_i$: vector of mass flows of all process chemicals for process i
$\hat{g}$: vector of GWP per unit mass of all process chemicals
g_e^{prod} : GWP of use phase electricity used in production

Energy and emissions due to transportation are given in (4.10) and (4.11) by e_{trans} and g_{trans}, where the two legs of transportation refer to transport from the fab to assembly plant, and from assembly site to use. The terms e_{trans} and g_{trans} are derived from transport distances d for each leg i, where t denotes truck; r, rail; b, boat and a, air freight, and the product and packaging mass for leg i, m_i.

$$e_{trans} = \frac{1}{Y_{wafer} Y_{line}\, n_{die}} \left[\sum_{i=1}^{2} \left(d_{i,t}\, e_t^d\, m_i + d_{i,r}\, e_r^d\, m_l + d_{i,b}\, e_b^d\, m_l + d_{i,air}\, e_{air}^d\, m_l \right) \right]$$
(4.10)

$$g_{trans} = \frac{1}{Y_{wafer} Y_{line}\, n_{die}} \left[\sum_{i=1}^{2} \left(d_{i,t}\, g_t^d\, m_i + d_{i,r}\, g_r^d\, m_l + d_{i,b}\, g_b^d\, m_l + d_{i,air}\, g_{air}^d\, m_l \right) \right]$$
(4.11)

e_t^d: energy use per mass transported unit distance
g_t^d: GWP per mass transported unit distance
m_i: product and packaging mass over leg i
$d_{i,x}$: distance over leg i transported by mode x

In (4.12), the use-phase energy consumption is found as the product of device power (p_{use}) and lifespan (t_{life}). Equation (4.13) describes use-phase GWP emissions as the GWP intensity of use-phase electricity (g_e^{use}) applied to use-phase energy consumption.

$$e_{use} = p_{use}\, t_{life}$$
(4.12)

$$g_{use} = g_e^{use}\, e_{use}$$
(4.13)

4.2.2 "Upstream" Materials

4.2.2.1 Chemicals

Among the life-cycle impacts of semiconductor products, the importance of energy-related emissions from the production of high purity chemicals has been noted previously [74, 93, 125, 130]. The limited LCA data available for exotic and/or high purity semiconductor process chemicals remains a challenge in quantifying these impacts. The production processes and formulas for advanced semiconductor processing materials are closely held intellectual property. Chemical textbooks and handbooks simply do not contain information about the production processes used to make them, and it is nearly impossible to identify the dominant production method among patent filings, as enterprises will at times file multiple patents describing different production pathways, or describe production recipes broadly so as to obfuscate the preferred method. While LCA data are available for some

Table 4.2 Chemical LCA data sources, part 1: process data

Chemical	Energy intensity MJ/kg	Carbon intensity gCO_2eq/g	Source
Al	260	22	[4]
Ar	3.6	0.31	[4]
C_2F_4	20	1.7	[127]
CH_4	38	3.3	[85]
CO	0.52	0.04	[45]
Cu	4.7	0.40	[10]
F_2	61	5.3	[41]
H_2	8.5	67	[4]
H_2O_2	12	1.0	[4]
H_2SO_4	0.040	0.00	[57]
HCl	0.91	0.08	[52]
He	0.83	0.07	[50]
HF (gas)	18	1.5	[98]
HF (liquid)	18	1.5	[4]
N_2	0.66	0.06	[55]
NF_3	40	3.4	[24]
NH_3	31	2.7	[1]
NH_4OH	15	1.3	[1]
O_2	1.8	0.15	[55]
Pb	2.0	0.2	[7]

basic chemicals used in wafer manufacturing, such as elemental gases, metals and common acids, it is usually representative of the industrial grade, with a purity of 99% or lower, rather than ultra-high purity or semiconductor grade (99.9997% to 99.9999999% pure).

This study uses a method of LCA data collection by which data based on process descriptions are used where available, and data from the Carnegie Mellon EIO-LCA database are used where costs are known. When no process LCA data and no cost information is known, an estimate for the energy intensity of chemical manufacturing developed by Overcash is used [66]. In this study, the "pharmaceuticals and medicines" rather than "photographic film and chemicals" commodity sector (NAICS #325400) is used in the EIO analysis for those materials which are high value specialty chemicals (those with a purchase price over $1,000 per kg), since the economic value of these materials is represented more closely by the former sector. The organic chemicals (NAICS #325190) and inorganic chemicals (NAICS #325180) commodities are used for the remaining materials, as appropriate. Although additional impact categories are available for those materials analyzed using EIO-LCA, the inventory is limited to primary energy demand and the GWP of emissions. Data sources for all in Tables 4.2–4.5.

The uncertainty of process data from textbooks and manuals is assumed to be zero, because it is unknown but assumed to be small as compared with other

Table 4.3 Chemical LCA data sources, part 2: process data

Chemical	Energy intensity MJ/kg	Carbon intensity gCO$_2$eq/g	Source
polyamides	115.0	9.9	[7]
Pt	270	23	[26]
SiH$_4$	2321	200	[43]
Sn	122	11	[7]
Ti	140	12	[1]
utility N$_2$	0.02	0.06	[55]

Table 4.4 Chemical LCA data sources, part 3: EIO-LCA data

Chemical	Energy intensity MJ/kg	Carbon intensity gCO$_2$eq/g
1,1-dichloro-1-fluoroethane	17	1.4
AsH$_3$	6.2E + 04	5.2E + 03
BCl$_3$	4.0	0.35
benzotriazole	17	1.4
bis tertiary-butylamino silane	5.9E + 04	4.9E + 03
C$_2$F$_6$	1.4E + 03	120
C$_4$F$_6$	1.3	0.11
C$_4$F$_8$	0.8	0.07
CF$_4$	1.0E + 03	86
CHF$_3$	59	5.1
Cl$_2$	1.3	0.11
CMP polishing solution	17	1.4
CuS silica slurry	17	1.4
DCS	5.3	0.45
HCl (gas)	0.7	0.06
NH$_4$OH	76	6.6
PH$_3$	1.9E + 05	1.6E + 04
SiCl$_4$	1.5E + 03	130
SiF$_4$	3.3	0.29
SiH$_4$	2.3E + 03	200
surfactant solution	17	1.4
TDMAT	5.5E + 04	4.6E + 03
TEOS	1.3E + 03	100
TMS	2.8E + 04	2.3E + 03

chemical LCA data sources. All data sources and impact values for materials using published process energy data are given in Table 4.2.

While EIO-LCA results for toxic releases or other impacts may have a lower precision, an uncertainty range of ±10% is assumed for EIO-LCA energy consumption and ±20% for GWG emissions based on the sources used by the CMU model for energy data. Chemicals using EIO-LCA data are given in Table 4.4.

Table 4.5 Chemical LCA data sources, part 3: process-based common value [66]

Energy intensity MJ/kg	Carbon intensity gCO$_2$eq/g	
3.1	0.26	
ArH	ethyl lactate	O$_3$
As	Fe$_2$O$_3$	OMCTS
Au	formaldehyde (CH$_2$O)	oxide CMP slurry
B$_2$H$_6$	GeH$_4$	p-cresol
BF$_3$	H$_3$PO$_4$	PDMAT
Br$_2$	HBr	PGME
C$_2$H$_2$	HCOOH	PGMEA
C$_2$H$_4$	HMDS	polyimide laminate
C$_2$H5OH	laminate solvent	Sn
citric acid	m-cresol	SO$_2$
CMP abrasive	MMA	Ta
Cr	N$_2$O	TDEAH
CuCl$_2$	Na$_2$B$_4$O7	TDMAS
CuSO$_4$	Ni	TMAH
DEA	n-methyl-2-pyrollidone	W
DMA	NO	W CMP slurry
	NO$_2$	WF$_6$

Data from the Kim/Overcash study has an uncertainty of +25%/−75% as described in their analysis [66]. The list of chemicals using this common value for energy and GWP per mass is given in Table 4.5.

4.2.2.2 Silicon

Silicon is the purest substance used among all semiconductor process materials. There are several processing steps that raw silica takes to become a pure silicon wafer, the substrate of semiconductor devices. Raw silica is refined into metallurgical grade silicon, which is twice refined to produce a single crystal ingot that is then sliced into wafers. The high embedded energy of the final product (approx. 2,000 KWh/kg) is due not only to the energy intensity of these processes, but also a cumulative low yield caused by the losses at each step. Full descriptions of the energy requirements and environmental emissions of high purity silicon production are available from previous sources [88, 130]. The LCA data provided by Williams and used in this study [130] is duplicated here for clarity (Table 4.6). This yield losses at each stage of production contribute to the high energy intensity of this material.

Table 4.6 Energy intensity of silicon production

Process step	Electrical energy/kg Si out (KWh)	Si yield (%)
Refining silica to mg-Si	13	90%
mg-Si to trichlorosilane	50	90%
Trichlorosilane to polysilicon	250	42%
Crystallization of polysilicon to sc-Si ingot	250	50%
Sawing sc-Si ingot to Si wafer	240	56%
Process chain from silica to wafers	2127	9.5%

Table 4.7 Global warming intensity of Santa Clara water

	Local supply	Imported	Recycled
Contribution of source	45%	51%	4%
kWh/l	0.0021	0.0019	0.0002

4.2.2.3   Water

The environmental impacts associated with the Santa Clara water supply are modeled using information from the Santa Clara Valley Water District and previous work on LCA of California water supplies by Stokes [118].

The Santa Clara Valley Water District infrastructure is composed of 3 treatment plants for local and imported water, one recycled water treatment facility, 142 miles of pipelines and 3 pumping stations. According to a report from the district board, approximately 51% of the water used in Santa Clara is imported, while 45% comes from local sources and the remaining 4% from recycled stocks [131]. (A regional desalination project is planned for construction; however, no water is desalinated by the Santa Clara water district at the time of writing.) Most water imported to Santa Clara comes from the Sacramento-San Joaquin River Delta via the South Bay Aqueduct, though a small fraction also comes from the Hetch-Hetchy reservoir via the San Francisco water system. Local water sources include groundwater basins and 10 surface reservoirs.

The life-cycle environmental impacts evaluated by Stokes for imported and recycled water from the Oceanside Water District in San Diego are applied, on a per volume basis, to the imported and recycled fractions of water in the Santa Clara system. Life-cycle environmental impacts associated with Santa Clara's locally sourced water are estimated based on the energy required for treatment and distribution of imported water in Stokes' model of Marin's water treatment works. The global warming emissions intensity for the power utility in Santa Clara (Pacific Gas and Electric), 280 $gCO_2eq./kWh$, is used. The energy intensity and percent contribution of each source is presented in Table 4.7. The resulting global warming emissions per liter of water provided in Santa Clara is 0.6 gCO_2eq.

4.2.3 Infrastructure and Equipment

The energy use and GWP for infrastructure and equipment are evaluated using EIO-LCA . Rock's Law is used to estimate the total cost of the fabrication facility and the costs of wafer fabrication equipment are taken as 70% of the total cost of the fab, based on a commonly stated approximation. Expenditures are depreciated over a 10 year period, using a straight line schedule, yielding an annual cost which is corrected to 1997 dollar values using the average U.S. inflation rate over the 1995–2008 period of 2.7%. Total costs for the building and equipment for each technology node are provided in Table 4.8.

4.2.4 Electricity

The emissions associated with electricity use at the different geographical locations of each life-cycle stage are reflected in the model. In the fabrication and use stages, emissions factors for electricity are specific to California, while the stages of chemical and infrastructure production are represented by each US industry average GWP emissions factors, via EIO-LCA [18].

4.2.4.1 GWP of Electricity

The environmental impacts associated with electricity supplied to the California plant are evaluated using two previous LCA of electricity generation, data from the EPA and information from Santa Clara's electric utility, Pacific Gas and Electric.

The electricity mix of Pacific Gas and Electric in 2008 was 47% natural gas, 23% nuclear, 13% large-scale hydroelectric, 4% coal, 4% biomass or other waste combustion, 4% geothermal, 3% small-scale hydroelectric, 2% wind and 0.1% solar photovoltaic [87]. The life-cycle GHG emission factors (g CO_2eq./kWh) for natural gas, coal, large scale hydroelectric and solar photovoltaic power are taken from the work of Pacca and Horvath [84], while that for nuclear electricity is taken from a study by Fthenakis [44] (Table 4.9). Direct GHG emissions for geothermal and biomass combustion are taken from the EPA [38]. Small hydro is considered to have the same impacts as large hydro. A national average for the Chinese grid of 877 gCO_2eq/kWh, based on a previous LCA [29], is used for the production scenario in China.

4.2.4.2 Primary Energy Use in Electricity Generation

In order to facilitate comparison with preceding studies, for most life-cycle stages, the convention of 10.7 MJ of primary energy per kWh electricity is used.

Table 4.8 Cost of Fab infrastructure and equipment

Year		1995	1998	1999	2001	2004	2007	2010	2013
Tech. node		350	250	180	130	90	65	45	32
Equipment cost per year of operation									
	$M	42	70.6353	84	118.794	199.787	336	565.082	672
	$M in 1997$	44.4	68.7	79.5	106	165	256	396	434
GWP	MTCO$_2$E	23025	35670	41274	55260	85610	132628	205469	225083
Energy	TJ	283	439	508	680	1054	1633	2530	2771
Building cost per year of operation									
	$B	0.018	0.030	0.036	0.051	0.086	0.144	0.204	0.242
	$M in 1997$	19	29	34	46	71	110	143	156
GWP	MTCO$_2$E	10514	16289	18847	25234	39094	60564	78899	86430
Energy	TJ	125	194	225	301	467	723	942	1032
Equipment impacts per wafer									
GWP	kgCO$_2$eq	53	83	96	128	198	307	476	521
Energy	MJ	656	1017	1176	1575	2440	3780	5856	6415
Infrastructure impacts per wafer									
GWP	kgCO$_2$eq	24	38	44	58	90	140	183	200
Energy	MJ	290	450	521	697	1080	1673	2180	2388
Totals per wafer									
tech node		350	250	180	130	90	65	45	32
GWP	kgCO$_2$eq	78	120	139	186	289	447	658	721
Energy	MJ	947	1467	1697	2272	3520	5453	8036	8803

Table 4.9 GWP intensity of electricity

Electricity mix		Carbon intensity gCO_2eq/kWh	Source
Coal	4%	811	Horvath, Pacca [84]
Nat Gas	47%	450	Horvath, Pacca
Nuclear	23%	25	Fthenakis [44]
Large Hydro	13%	41	Horvath, Pacca
Biomass/waste	4%	0	EPA [38]
Geothermal	4%	35	EPA
Small Hydro	3%	41	Horvath, Pacca
Wind	2%	7	Horvath, Pacca
Solar	0.1%	90	Horvath, Pacca

Table 4.10 Electricity generation by type, world average vs. California

	Conventional thermal	Hydro.	Nuclear	Geothermal, solar, wind, and waste/biomass
World average	69%	19%	9%	3%
California (PG&E)	52%	16%	23%	10%

This represents a worldwide average value for fuel consumption in electricity production [130]. The primary energy intensity of electricity supplied in Santa Clara is not documented, and since there have been no studies which provide net fuel intensity of nuclear, geothermal, wind or the other non-combustion generation technologies used by the California grid, the fuel intensity of the electricity used in fabrication is taken as the this worldwide average. In actuality, the primary energy intensity of Santa Clara electricity is estimated as the world average. A comparison of the contribution of each generation type is given in Table 4.10. Since most of the thermal generation in California is combined cycle natural gas combustion, and the contribution of renewables and nuclear are higher than the world average, the net primary energy demand for electricity production is somewhat lower than 10.7 MJ/kWh. For the purposes of this study, however, the global average is used.

The fuel intensity of electricity in China, however, is higher, with an average value of 12 MJ per kWh of electricity, due to an average lower conversion efficiency of power plants as well as higher losses in transmission and distribution [29].

4.2.5 Semiconductor Manufacturing

In this analysis the primary model for wafer manufacturing is located in Santa Clara, California, in the U.S. A separate scenario for production in China is developed in order to demonstrate the environmental effects of using China's electricity supply mix and neglecting per-fluorinated compound (PFC) abatement. Although PFC emissions may be abated in some fabs in China, the assumption is made that there are no controls on PFC emissions at the Chinese production site.

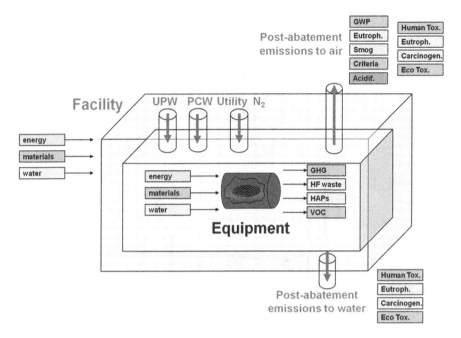

Fig. 4.2 Overview of mass and energy flows considered in the fab model

The mass and material flows are accounted at the level of the fab and equipment (Fig. 4.2).

4.2.5.1 Process Flows

A summary of changes to the process flow for each device is given in Table 4.11. The process change which has allowed the greatest reduction in GWP from one technology node to the next is the switch from in-situ plasma generation to remote plasma generation for etch and post-dielectric deposition chamber cleaning (Tables 4.11 and 4.12). (The impact of this particular technology change is not described in further detail in this thesis as it has been well-documented in previous literature [97].)

4.2.5.2 Facility and Process Equipment Energy Demand

While device design, process complexity and the length of the process flow grow relentlessly, total fab energy consumption has not increased at the same pace, and has at times decreased in the past decade due in large part to facility efficiency improvements. These changes are reflected in the model; At each technology node,

Table 4.11 Summary of process changes for each technology generation, part 1

Node (nm)	350	250	180	130
Wafer size (mm)	200	200	300	300
Interconnect	4 layer Al	5 layer Al	6 layer Al	6 layer Cu, Ta barrier
Starting wafer				SOI
Dielectric	USG PMD, USG ILD M1-M4	USG PMD, USG ILD M1-M5	FSG PMD, FSG ILD (Remote Clean) M1-M3, USG M4-M6	FSG PMD, FSG ILD (Remote Clean) M1-M3, USG M4-M6 (Remote Clean)
Contact	Ti silicide	Ti silicide	Ti silicide	Co Silicide
Strain Engineering				
Gate	RTO gate oxide	RTO gate oxide	RTO gate oxide	RTO gate oxide
Other	Nitride spacer for LDD	Nitride spacer for LDD	Nitride spacer for LDD	Nitride spacer for LDD
PR Strip	SPM – wet PR strip	SPM – wet PR strip	SPM – wet PR strip	dry plasma PR strip

Table 4.12 Summary of process changes for each technology generation, part 2

Node (nm)	90	65	45	32
Wafer size (mm)	300	300	300	300
Interconnect	7 layer Cu, Ta barrier	8 layer Cu, Ta barrier	9 layer Cu, Ta barrier	10 layer Cu, TaN barrier
Starting wafer	SOI	SOI	SOI	SOI
Dielectric	FSG PMD, BD with oxideBloK ILD M1-M3, USG M4-M7 (remote clean)	TEOS HARP PMD, BD with BloK ILD M1-M3, USG M4-M7 (remote clean)	TEOS HARP PMD, BD with BloK ILD M1-M4, USG M5-M9 (remote clean)	TEOS HARP PMD, BD with BloK ILD M1-M4, USG M5-M9 (remote clean)
Contact	Co Silicide	Ni Silicide, dry pre-clean	Ni Silicide, dry pre-clean	Ni Silicide, dry pre-clean
Strain Engineering	Nitride Cap, Spacer	Nitride Cap, Spacer nitridation of oxide: ONO gate stack	Epi SiGe, Nitride Cap	Epi SiGe, Nitride Cap
Gate	RTO gate oxide		ALD high-k dielectric (HfOx)	ALD high-k dielectric (HfOx)
PR Strip	dry plasma PR strip	dry plasma PR strip	dry plasma PR strip	dry plasma PR strip
Other	Source-Drain extension implant	Source-Drain extension implant	Source-Drain extension implant	Source-Drain extension implant
	Nitride spacer for LDD	Nitride spacer for LDD	Nitride spacer for LDD	Nitride spacer for LDD

improvements are made to certain facility equipment, such as the water chillers or exhaust pumps, which allow reduced energy consumption.

Rising energy costs as well as pressure to achieve GHG emission reduction goals set by the World Semiconductor Council, have driven fabs to reduce their total energy consumption. These efforts are reflected in the industry goals set in the ITRS, which show an ongoing effort to reduce facility energy consumption to between 0.5 and 0.7 kWh per cm^2 wafer area in the 1999–2005 time period [103, 105–107]. The trend may also be verified using an EIO perspective. By normalizing per unit of silicon area used, rather than by economic value of production, energy consumption can be analyzed independent from increases in off-shoring and outsourcing of fabrication by US companies or the increasing economic value of products. U.S. Census data from 1995 to 2005 show that the total electricity consumed by the semiconductor industry in the U.S., when normalized per area of silicon consumed by the industry, did not increase significantly from 1995 to 2005 [2,129]. The energy consumption per area of silicon consumed increases and decreases slightly over time, but was roughly the same in 2005 as in 1995, approximately 1.5 kWh/cm^2 [2, 130].

Energy efficiency goals have largely been achieved through changes to fab facility systems. Throughout the industry, improvements have been made to the energy efficiency of nearly all of the major fab systems: water cooling, exhaust flow, water distribution, clean room airflow, clean dry air (CDA) and facility nitrogen delivery systems, and chamber vacuum pumps. Facility energy efficiency improvements can be classified as advancements in both the technologies and in the techniques applied in fab design and operation. Higher efficiency pumps and fans, variable speed drives and improvements in ducting and clean room airflow arrangement such as mini-environments represent technological developments. Reduction of pressures in CDA and exhaust systems, optimization of clean room temperature and air speed and the use of larger of cooling towers to allow reduced chiller size are examples of operational improvements.

These advancements in facility system efficiencies are reflected in the fab and fab utilities models for each technology node in this study. The facility utility systems accounted for in this LCI model are: regular and hot ultra-pure water delivery, process cooling water, clean dry air delivery, and clean-room airflow and conditioning (commonly referred to as heating, ventilation and air conditioning, or HVAC).

Ultrapure Water (UPW): Ultra-pure water is produced in the fab through continuous cycling of water through a sequence of filter chambers and a reverse osmosis system. Typically, about 14 Watt-hours of electricity are required to produce one liter of UPW [83].

Process Cooling Water (PCW): Process cooling water is recycled upon return from the process equipment to be cooled using a chiller and cleaned, usually through ion exchange A typical PCW system delivers approximately 6,000 gallons per minute, and requires 0.91 Watt-hours per liter of PCW [83].

Table 4.13 Facility system changes by technology node

	Technology node (nm)
Increased CDA system pressure for advanced lithography	250
HVAC: Reduce ACR in cleanroom HVAC	180
House Scrubber: Use high efficiency VFD exhaust pumps and reduce pressure drop to scrubber	180
Increased sizing of cooling towers to allow reduced size of chillers	180
New PCW chiller with VSD	180
All facility system capacities are resized for 300 mm wafer fab	130
HVAC: Mini-environments, using Fan Filter Units with VFD	130
HVAC: Reduce fan sizes via redesign of air handling system	130

Clean Dry Air (CDA): The clean dry air system supplies contaminant free air above atmospheric pressure for use throughout the fab, for drying, pneumatic controls and other purposes. A typical system would consume approximately 3 Watt-hours per cubic foot of CDA, for compression, filtration, heating and cooling [83].

Heating Ventilation and Air Conditioning (HVAC): The air in the clean room must be maintained at very high levels of purity under positive pressure. Cleanroom air is maintained by continuous recirculation, monitoring and conditioning through a large installation of fans and filters, as well as cooling and humidity conditioning systems. These systems together constitute the HVAC system which consume a large fraction of the overall energy expenditure of the facility. In a typical fab, the make-up-air systems demand about 400 kW of electricity continuously [83].

A summary of these changes is given in Table 4.13. At the 250 nm node, the pressure maintained in the CDA delivery system is increased to support stepper systems required for this generation's photolithography tools. (This change does not enhance energy efficiency but was necessary to enable pneumatic stepping for lithography.) At the 180 nm node, the air change-over rate (ACR) is reduced in the clean room heating ventilation and air conditioning (HVAC) system, allowing fans speed to be lowered, the scrubber exhaust pumps are upgraded, a smaller and more efficient chiller, using a variable speed drive (VSD) is installed; Chiller use is also reduced by increasing the size of the cooling towers. Total facility energy consumption is cross-verified against industry reports and published literature [59, 104].

The wafer yield (good chips per wafer), line yield (finished wafers per wafer starts) and chip size are key variables which influence the environmental impacts per chip, as described in the Sensitivity section (Sect. 4.3.2). The values for these parameters at each technology node are based on industry average data (Table 4.14) [104].

Power data for process tools are based on measurements taken using three phase power measurement equipment, which have a maximum error of ±2.6%. Power requirements for facility systems are determined using mass flow analysis and facility energy consumption models, which are developed based on data

Table 4.14 Yields and chip sizes for each technology generation

Technology node (nm)	350	250	180	130	90	65	45	32
Line yield (finished wafer/wafer start)	58%	68%	73%	83%	83%	88%	88%	88%
Gross yield (chips/wafer)	117	201	249	429	429	463	463	463
Net yield (good chips/wafer)	88	151	187	322	322	347	347	347
Chip size (mm^2)	196	150	125	140	140	140	140	140

from industry and technical reports [75, 83]. Power and facilities requirements for process tools are from process equipment measurements [71] and requirements for abatement equipment requirements are based on manufacturers' specifications, which have an undefined error.

4.2.5.3 Process Emissions

The abatement of some PFC emissions are regulated by the Kyoto Protocol (in Annex I and II nations) and, in 1999, the World Semiconductor Council (WSC), which includes the semiconductor industry associations of Japan, Europe, Korea, Taiwan and the United States, issued a position paper which committed members to PFC emissions reduction by 10% of 1995 or 1999 baseline levels by the end of 2010. However, more than half of semiconductor production occurs outside of Kyoto Protocol Annex I and II nations, and, in 2008, almost 20% of semiconductor production capacity was held in China, Singapore and Malaysia, where the industrial consortia have not joined in the WSC. Thus, although PFC emissions may be abated in some fabs in China, the assumption is made that there are no controls on PFC emissions at the Chinese production site.

GWG emissions from each process step have been determined, pre- and post-abatement, using in-situ mass spectrometry and FT-IR analysis by a procedure which requires mass balance to be closed within 10% of chamber inputs. Each of these measurements thus has a maximum uncertainty of $\pm10\%$ for each element. For most materials, the uncertainty of the total mass of emissions per finished wafer can be considered as a uniform distribution with variance equal to $(10\%)^2$ of the expected value. For NF_3 which is at more than 30 points during processing of a single wafer the uncertainty is reduced via the central limit theorem, and the total mass flow is modeled as a normal distribution with variance equal to $(3.3\%)^2$ of the expected value. Global warming potentials are taken from [63].

Although these mass measurements themselves have low uncertainty, they should be considered a lower bound for direct emissions in fabrication. These measurements represent ideal conditions for PFC abatement. In addition, this model represents a fab where (point-of-use) POU PFC abatement equipment is installed on all PFC-using processes, while many fabs only use this type of abatement on some sets of processes. Due to both of these factors, an average case of fab PFC emissions would fall above the ideal cases illustrated in this chapter.

Table 4.15 GWP intensity of transportation

	Distance, fab. to assembly (miles)	Distance, assembly to use (miles)	CO_2 intensity (gCO_2/ton-mile)	Energy intensity (MJ/ton-mile)
Truck	50	200	187	2.7
Air freight	3000	3000	18	0.38

4.2.6   Transportation

Chips are typically cut and packaged at a facility separate from the wafer fabrication site, often in a different country or on a separate continent altogether [16]. Semiconductor products therefore travel twice within the production phase: wafers are transported from the fab to an assembly plant, where they are cut into die, packaged into chips and tested and finished chips are then transported to the place of eventual use.

The global industry of semiconductor packaging and testing, or "back-end" processing, is clustered in Vietnam, Malaysia, Costa Rica, Puerto Rico, China and the Philippines. Costa Rica is the closest location to Santa Clara and is therefore the location of assembly designated in this study.

Travel from the wafer fab to the assembly facility is taken as 50 miles by truck and 3,000 miles by plane, and from assembly to the final point of use, travel is 3,000 miles by air and 200 miles by truck. Energy consumption and GWP of emissions for truck and air freight are from Facanha [40]. The distance of each travel leg and its corresponding GWP impact and energy intensity is given in Table 4.15.

It is assumed that between wafer production and assembly, the finished wafer is transported in a wafer carrier and additional casing with a total weight of 500 g per 200 mm wafer or 700 g per 300 mm wafer. Between assembly and use, the product and packaging has an assumed weight of 20 g regardless of technology node. The total energy and GWP intensity of transport for each technology node is presented in Table 4.16.

4.2.7   Use Phase

The use phase represents the power consumption of the chip assuming a power supply efficiency of 70%. The lifetime of the chip is taken to be 6,000 h (in operation 8 h a day, 5 days per week, 50 weeks per year for 3 years) in a 17% active state, representing a business user. The activity rate of 17% is based on the SYSMARK standard [60, 89]. An assumption of 3 years is consistent with the literature, which identifies the typical lifespan of personal computers as 2–3 years in business applications and 4–5 years in residential use [25, 101, 129]. The lifetime assumed in this study would also be equivalent to an 18-month lifespan of a data center processor, operating continuously, with 95% uptime, at a 30% activity rate.

Table 4.16 Transportation energy and CO_2 emissions by technology node

Technology node (nm)	350	250	180	130	90	65	45	32
Wafer and carrier Weight (g)	646	646	1029	1029	1029	1029	1029	1029
Net die per wafer	88	151	187	322	322	347	443	347
Transported mass, fab. to asm. (g/die)	7.4	4.3	5.5	3.2	3.2	3.0	2.3	3.0
Transported mass, asm. to use (g/die)	20	20	20	20	20	20	20	21
CO_2eq, fab. to asm. (g/die)	33	19	24	14	14	13	10	13
CO_2eq, asm. to use (g/die)	89	89	89	89	89	89	89	89
Total GWP (g CO_2eq/die)	122	108	114	103	103	102	99	102
Energy, fab. to asm. (kJ)	470	273	351	204	204	189	148	189
Energy, asm. to use (kJ)	1283	1283	1283	1283	1283	1283	1283	1347
Total energy (MJ/die)	1.8	1.6	1.6	1.5	1.5	1.5	1.4	1.5

Table 4.17 Use phase power by ITRS technology node

Technology node (nm)	350	250	180	130	90	65	45	32
Year	1995	1998	1999	2001	2004	2007	2010	2013
Maximum rated Power (W)	14	23	25	61	84	104	146	158

As listed in Table 4.17, the average power requirements for logic chips are taken from the 2001–2007 International Semiconductor Manufacturing Roadmap reports [105–108] and, for years previous, from manufacturer's specifications. These power values represent operation at full capacity, or at a 100% activity rate.

The industry-defined average maximum chip power demand has risen from 14 to over 150 watts over the past 18 years. This maximum power does not represent average power in use.

IC performance is inextricably linked to power demand. As transistor size falls, the number of transistors per chip area increases, allowing greater computational performance. With scaling, reduced gate thicknesses allow reductions in gate voltage, but also cause leakage current increases. While there are many process technologies used to offset these general constraints, the overall trend is apparent in the industry roadmap. It is important to note that hardware and software improvements to energy conservation such as active power management have done much to offset these increases, particularly over the time period of this study. Power requirements for logic chips are nevertheless the main cause of rising energy-related life-cycle impacts, as will be shown in the Results section.

In order to compare impacts on a common basis of operational performance, the metrics of MIPS is used, rather than clock speed or transistor density, as a common measure of computational capacity. Transistor density is not ideal as a computational power metric because while increased transistor density usually results in increased computational power, the relation is not necessarily proportional. Although clock speed, which is dependent on transistor density, has been used as a popular measure of a CPU performance, computational power is determined by the

Table 4.18 Computational performance

Technology Node (nm)	350	250	180	130	90	65	45	32
Year	1995	1998	1999	2001	2004	2007	2010	2013
Computational Capacity (Mtrans/die)	15	28	32	63	153	307	773	2,190
Computational Capacity (MIPS/die)	100	540	1,350	6,500	9,730	27,100	59,500	147,600

CPU's number of transistors, architecture, instruction set, cache size and memory speed as well as clock rate. Also, clock rate has leveled off in the past few CPU generations, as heat dissipation has limited its increases. The rate of instructions performed, usually denoted in million instructions per second (MIPS) accounts for both the speed and design of the chip but is still highly dependent on the instruction sequences used to define the metric. Though instruction rate falls short of providing a perfect description of a CPU's performance as processors with different instruction sets or architectures are not comparable, instruction rate is a more representative metric than clock rate, and is offered here as a means of comparison in addition to transistor density (Table 4.18).

4.3 Results and Discussion

As technology has progressed, life-cycle energy use and greenhouse gas emissions have in general been increasing per wafer and per die but decreasing when normalized by computational power. Figure 4.3 shows how total life-cycle energy demands per wafer, per die and per million transistors have changed over the period under study.

The increases in per-wafer and per-die life-cycle impacts have one dominant cause: the escalation of use-phase chip power. The growth in per-wafer impacts, however, is also due to the lengthening of the manufacturing process flow and concomitant expansion in manufacturing infrastructure and equipment, as shown in (Figs. 4.4 and 4.7). At each technology node, the complexity of device design has increased, and the number of process steps required to produce a finished wafer has escalated. In this model, for example, production of a finished wafer entails 147 process steps at the 350 nm node, while the process flow for a 45 nm device consists of a total of 251 process steps. Given improvements to fab facility systems, as described in Sect. 4.2.5.2, growth in manufacturing energy use per wafer has been driven in larger part by process equipment than by facility systems (Fig. 4.7). The lengthening of the process flow follows from increasingly detailed construction necessary to scale down the device's transistors as well as additional interconnect layers to wire them together.

Growth in manufacturing and materials-related impacts over time has been counteracted to some extent by shrinking die sizes, which allow more die to fit on

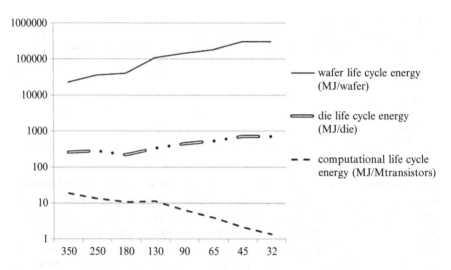

Fig. 4.3 Energy use per die, per wafer and per million transistors by technology node

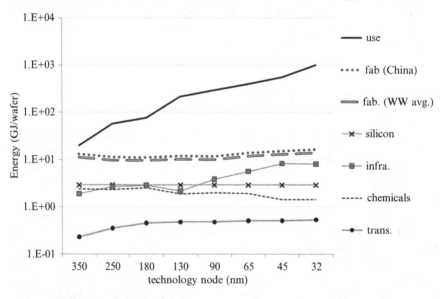

Fig. 4.4 Energy use per 300 mm wafer equivalent, by life-cycle stage, over eight technology nodes

each wafer. Thus, use-phase power is the primary reason for increases in impacts per die. For all technology generations, the use phase represents the largest proportion of energy-related impacts per die among the life-cycle phases. The dominance of the use phase has also increased over time, with use contributing about 40% of life-cycle GWP consumption per die at the 350 nm node, and over 95% per die at the 32 nm node. Despite the long distances that semiconductor wafers and chips are typically shipped by air during production and prior to use, GWP of transportation is almost insignificant due to the small mass of the product (Figs. 4.5 and 4.6).

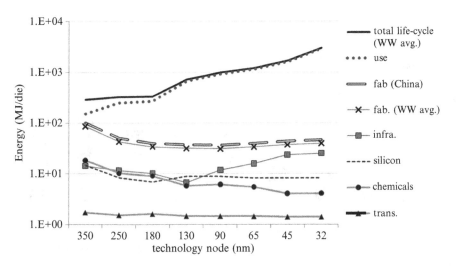

Fig. 4.5   Energy use per die, by life-cycle stage, over eight technology nodes

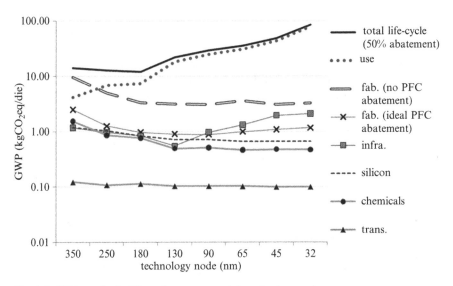

Fig. 4.6   GWP per die, by life-cycle stage, over eight technology nodes

GWP impacts are distributed over the life-cycle stages similarly to primary energy consumption, with a slightly larger fraction of impacts represented in the Chinese production phase due to the additional GWP of PFC emissions, as shown in Fig. 4.6. GWP of GHG emissions for the fab in China are much higher than those of the plant in California due to differences in electricity supply and the assumed lack of PFC abatement in the Chinese fab. Notably, NF_3 is not regulated by the Kyoto Protocol, but is among the PFCs GWG which are used in highest volume

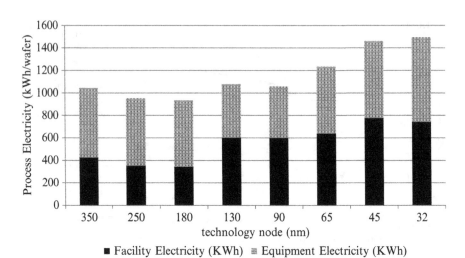

Fig. 4.7 Energy use per wafer by facility and process equipment, over eight technology nodes

in the semiconductor industry [56, 96]. Although NF_3 is largely broken down into F_2 during processing, a small amount (about 1%) leaves the chamber unreacted. In this model, at the most recent technology generation, NF_3 is emitted at a rate of 4.8 grams per wafer before abatement. For a fab with a capacity of 10,000 wafer-starts per week and no PFC abatement, this would result in GWP of close to 17,000 metric tons of CO_2 equivalent per year.

In spite of increases in energy use on a per-wafer basis (Fig. 4.7), the improvement of several production performance metrics has allowed reductions in the manufacturing energy and GWP per chip (Fig. 4.5). Line yield reflects wasted processing used for process monitoring, testing and wafer loss in the form of damage or breakage. Although wafer damage has remained the same over the years, at about 2%, the number of test or monitor wafers per finished wafer has been reduced over the last decade, resulting in higher average line yields [42, 58, 126, 136] (Table 4.14). Although reduced feature sizes have made maintaining wafer yield difficult, industry reports indicate that wafer yields for full scale production have not fallen with decreasing device dimensions. Mature wafer yield is assumed to be 75% for all technology nodes, based on ITRS reports [103, 105–107].

Over the period observed in this study, the computational power of an average CPU grows approximately exponentially, which results in a significant reduction in the environmental impact per computational unit of chip (Figs. 4.8 and 4.9). The question as to whether the appropriate functional unit is a single chip or 1 million transistors worth of chip is not clear-cut because as the computational capacity per area of finished wafer has increased, the computational demands of computing have also increased. (This topic of discussion is expanded in Chap. 7.)

As web-based applications and thin clients such as smart phones are used more frequently, computational load shifts to data centers on the internet and away from desktops and laptops. In this arrangement, data center servers can allocate memory

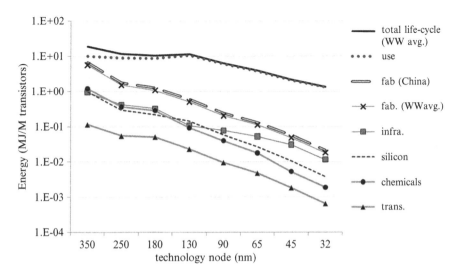

Fig. 4.8 Life-cycle energy use per computational power

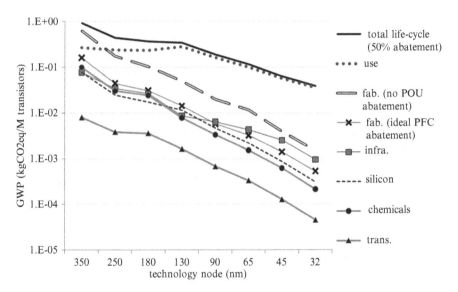

Fig. 4.9 GWP per computational power

and operational demands to reduce the number of active devices, such that the functional unit may more appropriately be a metric of computing power than a physical device. However, the additional demands of communication over a network also add to total energy consumption of internet-based applications. The question as to whether a specific case of network-based thin client computing would have a lower environmental impact than local desktop computing requires consideration of

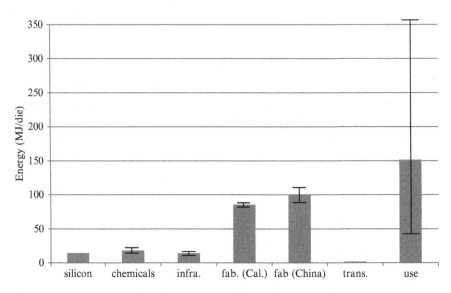

Fig. 4.10 Energy use per die by life-cycle phase, 350 nm node

additional factors concerning the application and network, however, and the results of this study should not be considered as favorable to either computing structure.

Results for energy use and GWP data per die for each life-cycle stage at every generation are provided in the Appendix. To determine impact values for a specific logic chip, the appropriate technology generation (e.g., 65 nm) and chip size should be used. (The average chip areas used in this study are listed in Table 4.14). Dual core and quad core CPUs, which are larger, will have higher impact values for all life-cycle stages before use. If the chip power is known, recalculate the use phase power, as the device's rated power is the most important variable in determining life-cycle energy demand. If the subject of the LCA is a computer or other electronic appliance, the efficiency of the power supply should be considered.

4.3.1 Uncertainty Assessment

The value of this or any LCA is wholly dependent on the quality, accuracy and precision of the underlying data. Results are presented here with their associated uncertainty to provide a more accurate representation of the possible range of impacts, and uncertainties are analyzed to evaluate their impact on the results. Best case, worst case and expected values of results are calculated from lower bound, upper bound and nominal model parameters. The energy and GWP of emissions at each life-cycle stage are presented with their uncertainty ranges for earliest technology generation, the 350 nm node in Figs. 4.10 and 4.11, the most recent 45 nm, Figs. 4.12 and 4.13.

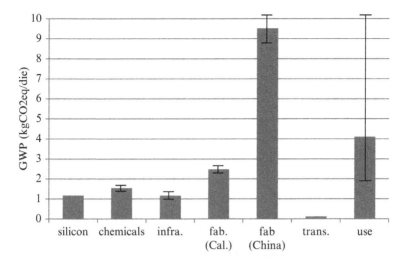

Fig. 4.11 GWP of emissions per die by life-cycle phase, 350 nm node

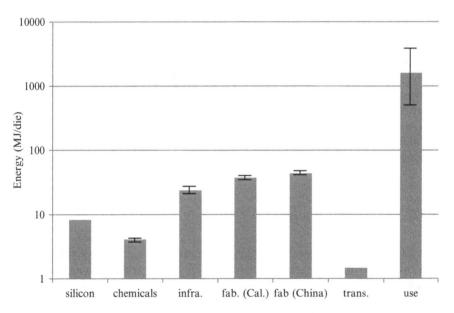

Fig. 4.12 Energy use per die by life-cycle stage, 45 nm node. Note: log scale attenuates appearance of uncertainty

The life-cycle stage with the greatest data uncertainty is chemical production. Of the chemicals included in the model, 30 are represented by process data from textbooks and manuals, 27 are accounted for by using EIO-LCA and the remaining 53 are assigned a common value based on a previous study of chemical life-cycle

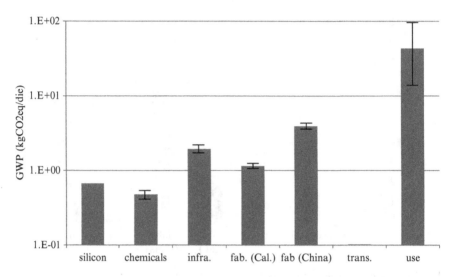

Fig. 4.13 GWP of emissions per die by life-cycle phase, 45 nm node. Note: log scale attenuates appearance of uncertainty

inventory modeling [66]. The data for the latter group has an uncertainty range of 75% below and 25% above the nominal value [66], which constitutes the largest contribution to uncertainty in the model's chemicals data. The chemicals with the greatest contribution to uncertainty differ for each technology generation, but the top contributors for all nodes include ammonia, CMP slurries and agents, phosphine (PH_3), carbon tetrafluoride (CF_4), hydrogen and silane (SiH_4). Life-cycle data for ultra high purity forms of these chemicals would have the greatest benefit to uncertainty reduction in future LCA of semiconductors.

4.3.2 Sensitivity Analysis

To evaluate the ramifications of changes to model parameters, a sensitivity analysis is performed. Given that the parameter values are limited within a finite range, and because the output of the model is monotonic over these values, the sensitivity of the model is evaluated using local methods via differential analysis [82]. The simple derivative is used to determine the absolute (S_x^E) and relative sensitivities ($\overline{S}_x^E$) of energy consumption and global warming potential to each model parameter (4.14) and (4.15). Energy and GWP impacts have the highest relative sensitivity to wafer yield, line yield and net die per wafer, followed by tool and facility utilization factors. Although the relative sensitivity of impacts to use-phase power is lower than to other parameters, processor power demand is the variable with the most absolute influence over life-cycle energy in all technology generations. The energy and GWP intensity of chemical production, which have the largest uncertainty of all model variables, are among the parameters to which impacts results are least sensitive.

$$\text{Absolute sensitivity:} \quad S_x^E = \frac{\delta E}{\delta x}\bigg|_o \tag{4.14}$$

$$\text{Relative sensitivity:} \quad \overline{S}_x^E = \frac{\delta E}{\delta x}\bigg|_o \frac{x_o}{E(x_o)} \tag{4.15}$$

4.3.2.1   Sensitivity of Energy Consumption

Of all production-related parameters, the total life-cycle energy per die is most sensitive to the line yield (Y_{line}), wafer yield (Y_{wafer}), and gross die per wafer (n_{die}), as shown in (4.16), (4.17) and (4.18). These relations reveal that the lower the original yield, the greater the influence a percentage change in yield improvement has on energy consumption. At the 350 nm node, for example, the line yield is 58% and the wafer yield is 75%, so improvement in line yield should be a higher priority than wafer yield. At the 45 nm node, however, the line yield is at 88% while the wafer yield is 75%, so an increase in the latter would have a greater effect.

$$\frac{\delta\,(e_{total})}{\delta Y_{line}} = -(Y_{line})^{-2}\,e_{subtot} \tag{4.16}$$

$$\frac{\delta\,(e_{total})}{\delta Y_{wafer}} = -(Y_{wafer})^{-2}\,e_{subtot} \tag{4.17}$$

$$\frac{\delta\,(e_{total})}{\delta n_{die}} = -(n_{die})^{-2}\,e_{subtot} \tag{4.18}$$

$$\text{where } e_{subtot} = \left[\sum_{j=1}^{m} e_j^\$ c_j \sum_{i=1}^{n} m_{ij} + \sum_{k=1}^{q} e_k^m \sum_{i=1}^{n} m_{ij}\right]$$

$$+ \left[\sum_{i=1}^{n} \frac{t_i\,p_i + \hat{e}_w^v\,\hat{w}_i\,t_i}{\varepsilon_i\,u} + \left(\frac{p_{hvac} + p_{lt} + p_{trt} + p_{cda} + p_{exh}}{c}\right)\right]$$

$$+ \left[\sum_{l=1}^{2} (d_t\,e_t^d\,m_l + d_r\,e_r^d\,m_l + d_b\,e_b^d\,m_l + d_{air}\,e_{air}^d\,m_l)\right]$$

The energy used in production alone is also equally sensitive to both the tool and process utilizations (the percentage time that the tool is on and the percentage of time that it is active while it is on) as to yield. Although the idle power is lower than active power, any time spent in idle mode results in wasted power. Equivalently, any decrease in tool utilization results in wasted power at the facility level, because fans, pumps and facility systems continue to operate during tool downtime. Because utilization factors affect only the production stage rather than all life-cycle stages, their influence on life-cycle impacts is weaker than that of the yield parameters.

$$\frac{\delta\,(e_{total})}{\delta\varepsilon_i} = \varepsilon_i^{-2}\left(\frac{1}{u\,Y_{wafer}Y_{line}\,n_{die}}\right)\left[\sum_{i=1}^{n} t_i\,p_i + \hat{e}_w^v\,\hat{w}_i\,t_i\right] \tag{4.19}$$

$$\frac{\delta\,(e_{total})}{\delta u} = u^{-2}\left(\frac{1}{\varepsilon_i\,Y_{wafer}Y_{line}\,n_{die}}\right)\left[\sum_{i=1}^{n} t_i\,p_i + \hat{e}_w^v\,\hat{w}_i\,t_i\right] \tag{4.20}$$

The capacity of the facility, i.e., the number of wafers produced per day or week, can also raise or lower the energy consumption, as shown in (4.21). However, capacity is limited by a number of factors. Production volume is governed by throughput, the number of manufacturing tools in the fab and the design of the production flow. The quantity of tools can be increased only up to a point before facility systems must be resized, or their efficacy suffers. In other words, capacity (c) is limited by the facility power consumption parameters p_hvac, p_lt, p_trt, p_cda and p_exh in a relationship that is described outside of this model.

$$\frac{\delta\,(e_{total})}{\delta c} = \frac{1}{c^2}\frac{1}{Y_{wafer}Y_{line}\,n_{die}}\,(p_{hvac} + p_{lt} + p_{trt} + p_{cda} + p_{exh}) \tag{4.21}$$

Changes in yield, utilization and capacity have a nonlinear effect on life-cycle energy and global warming emissions, and thus have high relative sensitivity values (4.23) while the use-phase power, the global warming intensity of electricity (at the locations of production and of use) and the power consumption of facility systems and individual tools all have a direct relationship (4.22), with varying degrees of influence.

$$\frac{\delta\,(e_{total})}{\delta p_{fac}} = \frac{1}{Y_{wafer}Y_{line}\,n_{die}}\,\left(p_{fac}\right) \tag{4.22}$$

$$\left|\frac{\delta\,(e_{total})}{\delta Y_{wafer}}\right|_{Y_{wafer}^{45nm}} = -\left(Y_{wafer}^{45nm}\right)^{-2} e_{subtot}^{45nm} \tag{4.23}$$

The rank of these variables according to their influence over life-cycle energy and GWP emissions differs for each technology node. However, impacts have the highest absolute sensitivity to use-phase power consumption at all technology nodes. (The fraction of life-cycle energy consumption taken by the use phase changes over the generations but remains high, as illustrated in the Results section.) At the 45 nm node, life-cycle energy consumption has the next highest absolute sensitivity to the primary energy intensity of use-phase electricity followed by the power consumption of wafer fabrication equipment and facility operations.

Table 4.19 Data quality assessment

	Reliability	Completeness	Temporal correlation	Geograph. correlation	Technical equival.
Chemicals (Process LCA)	2	5	5	2	2-3
Chemicals (EIO-LCA)	2	1	4	2	3
Process electricity (California mix)	1	1	1	1	1
All other electricity (world mix)	2	3	2	2	2
Wafer fabrication: atmospheric furnace and litho.	2	1	3	n/a	2
Wafer fabrication: all other processes	1	1	1	n/a	1
Point-of-use abatement	1	1	1	n/a	1
Facility abatement	2	2	2	n/a	1
Transportation	2-3	1	1	2	2
Use (chip power, performance)	1	2	1	1	1

4.3.2.2   Sensitivity of GWP Impacts

Life-cycle GWP, like energy, has the highest relative sensitivity to line and wafer yields, and net die per wafer. At the most recent node, GWP has the highest absolute sensitivity to the power consumption of electricity in the use phase, followed by the GWP intensity of use-phase electricity, the global warming intensity of transportation, the GWP of electricity used in wafer manufacturing and the energy consumption of wafer fabrication in descending order.

$$\frac{\delta\left(g_{total}\right)}{\delta g_e^{use}} = e_{use} \tag{4.24}$$

$$\frac{\delta\left(g_{total}\right)}{\delta g_x^d} = \frac{1}{Y_{wafer}Y_{line}\,n_{die}} \sum_{i=1}^{2} (d_{i,x}\ m_i) \tag{4.25}$$

$$\frac{\delta\left(g_{total}\right)}{\delta g_e^{prod}} = e_{prod} \tag{4.26}$$

4.3.3   Data Quality

A data quality assessment following the template of Weidema is provided in Table 4.19 [128]. The quality of data is high: all of the LCA data, with the exception of chemical and infrastructure data, come from sources that are specific to the process, geographical location and time period of the study.

4.4 Discussion

The results of this study enable LCA practitioners to answer important questions concerning the energy-related environmental impacts of computing with greater certainty than ever before. The life-cycle impacts for energy and GWP of semiconductor chips presented in this analysis are more complete, accurate and transparent than those of any previous study, and data are presented for chips spanning many generations, from 1995 to 2013. The quality of most data is high: all of the LCA data, with the exception of chemical and infrastructure data, are specific to the process, geographical location and time period of the study. The impacts determined through EIO for the fab infrastructure may be an overestimate, and the value expended in fab construction that is of a non-physical nature. Though life-cycle energy and emissions data for high purity chemicals would be useful for future semiconductor LCA, the uncertainty in chemical data is mitigated by the comparatively small contribution of upstream activity to total life-cycle energy and GWP. Energy and GWP impacts for semiconductor logic chips are clearly dominated by the use phase. Chip power demand and the GWP of use-phase electricity are thus the variables with the largest influence over energy-related life-cycle impacts. Production yield, die size, geographical location or electrical energy supply of the plant and the choice to abate PFCs are the most important metrics and decisions to be made concerning energy and GWP impacts in the production stage.

Chapter 5
Life-Cycle Assessment of CMOS Logic

5.1 Introduction

Determination of the life-cycle environmental and human health impacts of semiconductor logic is essential to a better understanding of the role information technology can play in achieving energy efficiency or global warming potential reduction goals. This chapter provides a life-cycle assessment for digital logic chips over 7 technology generations, spanning from 1995–2013. Environmental indicators include global warming potential, acidification, eutrophication, ground-level ozone (smog) formation, potential human cancer and non-cancer health effects, ecotoxicity and water use. While impacts per device area related to fabrication infrastructure and use-phase electricity and have increased steadily, those due to transportation and fabrication direct emissions have fallen as a result of changes in process technology, device and wafer sizes and yields over the generations. Electricity, particularly in the use phase, and direct emissions from fabrication are the most important contributors to life-cycle impacts. Despite the large quantities of water used in fabrication, water consumption is primarily driven by electricity generated for use-phase power. Reducing power consumption in the use phase is the most effective way to limit impacts, particularly for the more recent generations of logic.

The complementary metal oxide semiconductor (CMOS) transistor structure is the most common form of digital logic used in electronics today. This chapter presents a life-cycle assessment (LCA) for generic CMOS logic at each technology node over a 18 year period, from the 1995-era 350 nm node to the 32 nm node, which enters large scale production in 2013. The purpose of this chapter is to provide a detailed, complete, transparent and accurate inventory of the environmental impacts of many generations of logic chips in order to investigate trends in emissions over time and to allow LCA practitioners to more accurately model electronic equipment, as well as services enabled by electronics. Previous published work in the area of semiconductor LCA has included four environmental impact studies from industry [102, 119, 120, 134] which report impacts for wafer

S.B. Boyd, *Life-Cycle Assessment of Semiconductors*, DOI 10.1007/978-1-4419-9988-7_5,   69
© Springer Science+Business Media, LLC 2012

fabrication and, in some cases, also use and the production of materials. Most do not include impacts associated with the production of facility infrastructure or process chemicals (aka, "upstream" impacts). Possibly because these reports have all been conference papers supported by spoken presentations, they lack complete explanation of data collection methodologies and do not report complete inventory data. None of these studies mention the use of measurement to determine the mass of emissions from the fabrication facility but rather use estimation methods or do not explain whether or how they account for facility emissions. These studies also do not report data uncertainty or discuss the influence of data uncertainty on results. Several researchers have noted that the lack of LCA data for semiconductor devices is a stumbling block in LCA of electronics, and that there is particular need for more complete or transparent LCA of semiconductors [11,12,79,93,94,125,130]. Academic work related to semiconductor LCA includes a study from Murphy which presents a methodology for parametric semiconductor life cycle inventory (LCI) models based on process specifications [79]. Williams has reported energy consumption in logic manufacturing [27] and created an LCA of a memory chip, using both economic-level data and data provided from anonymous industrial contributors, and highlights the importance of upstream impacts and the need for more accurate LCI data for high purity chemicals [74, 129, 130]. Plepys also underscores the need for accounting of the life-cycle stages preceding wafer fabrication in semiconductor LCA [93, 94]. In an earlier paper, Plepys explored a rebound effect in ICT consumption, analogous to the rebound effect of dropping energy prices, whereby advancing technology incites the need for ever faster and more powerful ICT and counteracts the environmental efficiencies of technology advancement [91]. A more recent paper from Williams re-introduces this topic with an examination of different functional units in semiconductor LCA [27], and the discussion of functional unit choice is continued in this chapter.

In this study, the issues with industry-reported LCA studies described above are addressed. Material demands and emissions have been determined using Fourier-transform infrared and mass spectrometric measurement of process inputs, chamber emissions and post-point-of-use abatement emissions. Upstream impacts associated with fabrication facility (fab) infrastructure and process chemicals, as well as water supply are included. The uncertainty associated with each type of data is reported and the sensitivity of results to uncertainty and changes in model parameters is evaluated. Methodology, model assumptions, and inventory data are stated so that the study is transparent, reproducible and adaptable so as to be useful in downstream LCA of electronics. By presenting LCA data for many generations of logic, different types of electronics may be modeled. Chips at the 350 nm node (first produced in volume in 1995) are still currently used in embedded logic for appliances and toys, while an average new personal computer purchased today would contain 65 nm logic. The term "technology node" and the measurements of 350 to 32 nm refer to the half-width of the first interconnect layer associated with memory of a given technology generation, and are used as shorthand for relative transistor sizes [104].

5.2 Methods

These life-cycle inventories (LCIs) describe a production scenario with wafer manufacturing in Santa Clara, California, using chemicals, equipment and construction materials produced in the U.S. "Back-end" operations (die packaging and testing) are located 3000 miles away. The inventory is a hybrid model, containing primarily process data, supplemented by economic input-output LCA (EIO-LCA) data from the Carnegie Mellon database [18] where process data are unavailable. The functional unit of the study is one die over a lifetime of 6,000 h, though data are provided in the Supporting Information to reevaluate these results for a different use phase chip power, lifetime or die size and to allow normalization by computational power or number of transistors.

The functional unit drastically alters how the life-cycle impacts of semiconductors appear and there are arguments for every option of functional unit: an average device, a certain device area, a given number of transistors or another metric of computational power. The natural first choice is a measure of computational power, such as one million transistors, because this seems to reflect a constant functionality. However, the functionality of one million transistors has decreased over time, as the average personal computer has required increasing computational power over the past 18 years. Presenting impacts per million transistors shows dramatic decreases in impacts over time which do not match real-world dynamics [91, 15a] as, for example, one million transistors today do not provide the same functionality as a decade ago in personal computing applications. Because this study spans an eighteen year period, results reported only per million transistors would exhibit misleading trends. The functional unit of an average-sized personal computer (PC) central processor for each year reflects a set functionality over time because this unit serves the same product function within its corresponding timeframe. The functional unit is thus one average-sized die, as defined for cost-performance CMOS logic by the International Technology Roadmap for Semiconductors (ITRS) [104] but these results may also be adapted to represent any CMOS logic-based chip, if the chip size or number of transistors is known (Fig. 5.1).

The LCIs for wafer production are built on a set of process, device and fab spreadsheet models. Each process model represents one process step (e.g., chemical vapor deposition) with a set of energy and mass flows per wafer into and out of the manufacturing equipment, based on measurements taken at the process chamber inlet, chamber outlet and post point-of-use (POU) abatement. Each device model

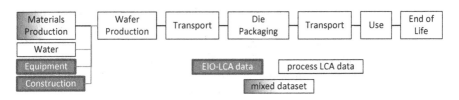

Fig. 5.1 Life-cycle stages with data source types

defines the device size, wafer size and typical yield for the device's technology generation as well as the process flow – the order and number of process steps used to make the device. Chip sizes and yield models are those developed through ITRS [104]. Each fab model represents all of the infrastructure and fab facility systems beyond the process tools and POU abatement equipment, which are characteristic to each technology node. The energy and resource demands for each fab model are based on the capacity of its facility systems, which change with wafer size, as well as the demands for utility nitrogen, process cooling water, industrial city water and abatement chemicals, as determined by the process flow at each technology node. The fab models also reflect technology and operational changes which have resulted in facility energy efficiency improvements over the last 18 years.

These LCIs represent the impacts associated with all life-cycle stages, though data for end-of-life effects of ICs are limited to the impacts of lead, which is present in some solders. Many other end-of-life impacts resulting from disposal of computers have been identified in the study of e-waste. A great deal of effort has been focused on the end-of-life of computer systems because irresponsible recycling practices can produce dramatic and visible human health and environmental impacts. The major pollutants associated with e-waste (flame retardants, polychlorinated biphenyls, dioxins/furans, polycyclic aromatic hydrocarbons, lead, cadmium and mercury) are largely emitted from the incineration or chemical breakdown of circuit boards, wiring, housing and displays. Although there may be harmful emissions from the decomposition or combustion of a logic chip, these have not yet been measured in isolation, but remain an important topic for future work. In this study, EOL impacts include only the lead emissions from wire-bonding solder contained inside the packaged chip. EOL lead emissions stop at 2006, when the European Union's Restriction on Hazardous Substances (RoHS) regulation banned lead-containing solders. Public statements indicate that major IC manufacturers had switched to lead-free solder for products shipped worldwide by 2006, even though RoHS only regulates products sold in Europe.

5.2.1 Manufacturing Process Power and Emissions

The mass flows for each process step, with the exception of lithography and certain thermal steps, have been determined using in-line mass spectrometry and Fourier transform infrared (FT-IR) spectroscopy. (Details of the process models can be found in a previous, related study of an individual chip [71].) Each emission measurement closes mass balance within 10% of chamber or POU abatement system inputs and thus has a maximum uncertainty of $\pm 10\%$. Process equipment power consumptions are based on measurements which have an associated error of $\pm 2.5\%$ [71]. Emissions and power consumption of photolithographic and thermal processes are taken from the process measurements from Murphy [79], supplemented by data from an unpublished academic report from Peterson [86].

5.2.2 Facility Utility Systems and Facility Energy Efficiency

The facility utility systems accounted for in this LCI model are: regular and hot ultra-pure water, process cooling water, clean dry air delivery, and clean-room airflow and conditioning (commonly referred to as heating, ventilation and air conditioning, or HVAC).

Ultrapure Water (UPW): Ultra-pure water is produced in the fab through continuous cycling of water through a sequence of filter chambers and a reverse osmosis system.

Process Cooling Water (PCW): Process cooling water is used to cool high-energy equipment and is recycled, cleaned and cooled. The PCW system therefore consists of using a chiller and filtration or ion exchange system.

Clean Dry Air (CDA): The clean dry air system supplies contaminant free air above atmospheric pressure for use throughout the fab, for drying, pneumatic controls and other purposes.

Heating Ventilation and Air Conditioning (HVAC): The air in the clean room must be maintained at very high levels of purity under positive pressure. Clean-room air is maintained by continuous recirculation, monitoring and conditioning through a large installation of fans and filters, as well as cooling and humidity conditioning systems. These systems together constitute the HVAC system which consume a large fraction of the overall energy expenditure of the facility.

The techniques used in industry to optimize the sizing and operation of fab utility systems include a long list of practices, including more efficient cleanroom airflow (including the use of mini-environments), reduced clean dry air (CDA) and nitrogen pressures, reduced exhaust system pressures and increased sizing of cooling towers to allow reduced chiller sizes. More information concerning the sizing and design of facility systems in the fab models is given in the previous chapter (Table 4.13).

5.2.3 Abatement

The largest and most fundamental of the facility abatement systems is the "house scrubber", an enclosed, water-sprayed matrix of inert mesh. This system captures gaseous wastes, as is specifically intended to capture inorganic, non-fluorinated emissions, largely acids. Once captured in the water spray, the aqueous mixture is sent as liquid effluent to the acid waste neutralization (AWN) system, which continuously monitors and corrects the pH of the incoming liquid waste.

Gaseous ammonia is emitted in small quantities from most nitride chemical vapor deposition (CVD) processes, either as unreacted precursor or as a byproduct emission. Fabs with gaseous ammonia exhaust are fitted with a separate ammonia exhaust system and scrubber in order to prevent particulate formation, clogging and corrosion in the acid exhaust system. Gaseous ammonia waste is captured using a water scrubber similar in design to the facility acid scrubber but about a tenth of the size [83].

CMOS logic fabs use large quantities of both liquid ammonia and sulfuric acid in wafer cleaning processes. Liquid ammonia, collected via drain, may be recycled on site using membrane filtration or distillation, or treated using sulfuric acid to produce ammonium sulfate. In this model, the latter is assumed and thus ammonium sulfate, which results from the neutralization of ammonia and sulfuric acid effluents in the AWN system, is among the liquid wastes produced in the highest volume by wafer fabrication in this model.

There are several combinations of treatment methods which may be used to address the liquid effluent of copper CMP processes. Copper CMP waste treatment is described in the work of Krishnan [70, 73] as a sequence of ion exchange, microfiltration, activated carbon filtering and filter pressing. An ion exchange resin bed removes copper and is regenerated at the fab using sulfuric acid, to produce CuSO4 liquid waste. Slurry particles are filtered and pressed into a solid non-hazardous waste which is sent to a landfill. The remaining water contains less than 2 ppm dissolved copper and is sent to the acid waste neutralization (AWN) system. The concentrated CuSO4 liquid is sent offsite as hazardous waste to be electrowinned for copper recovery or possibly purified into a useable byproduct.

The fluoride waste system treats fluoride wastewater using CaOH and a flocculant material to produce non-hazardous solids containing calcium fluorite (CaF_2).

CVD steps emitting PFCs require combustion and water scrubbing, plasma or cold bed adsorption point-of-use (POU) abatement because water scrubbing in the house scrubber will not break down these compounds (and in some cases may form hazardous, reactive fluorinated byproducts [110]). CVD steps emitting silane or hydrogen above flammable concentrations also require immediate combustion of their emissions in POU systems due to the risk of explosion in exhaust lines. Implant processes emitting phosphine and arsine are typically abated using cold bed adsorption systems at point-of-use.

In the model used in this study, both point of use (POU) and central facility abatement systems are included. The central abatement systems in each fab model include a central acid scrubber, an ammonia scrubber, a volatile organic compound (VOC) oxidizer and an acid waste neutralization unit as well as copper CMP and fluoride wastewater treatment systems. The abatement efficiencies of these central systems for gaseous acids, ammonia and VOCs and are based on measurements published by semiconductor industry members [19, 20, 76, 110, 111, 117]. Conversion of liquid waste is calculated based on expected reactions and the pH requirements of effluents to the public water treatment system. Combustion and water scrubbing, plasma oxidation and cold bed adsorption POU systems are associated with certain process steps. The abatement efficacies of POU PFC abatement units are taken from process-level pre- and post-POU abatement emissions measurements [71].

A summary of the process types requiring POU abatement systems is given in Table 5.1. CVD steps emitting PFCs require combustion and water scrubbing

Table 5.1 POU abatement

POU abatement type	Processes
Plasma	Nitride, oxide, advanced lo-k dielectric and undoped silicate glass (USG) etch
Burn and scrub	All other CVD and etch processes emitting PFCs or flammable gases
Cold bed adsorption	Implant

or plasma POU abatement because water scrubbing alone does not break down these compounds (and in some cases may form reactive fluorinated byproducts). CVD steps emitting silane or hydrogen above flammable concentrations also require immediate combustion of their emissions in POU systems due to the risk of explosion in exhaust lines. Implant processes emitting phosphine and arsine are abated using cold bed adsorption systems.

5.2.4 Environmental Impact Metrics

Environmental effects are characterized using the mid-point impact factors established in the Tool for the Reduction and Assessment of Chemical and other environmental Impacts (TRACI), a program developed by the US EPA. About a third of the TRACI impact factors used in the model are specific to California, and the remaining are representative of the United States in general. All photochemical smog formation impact factors are for California, and for some chemicals acidification, eutrophication, human health criteria and human non-cancer health impact factors are also California-specific. Impacts for ecotoxicity and carcinogenic human health effects are all US-general. Because impact factors specific to Costa Rica are not available, the same factors are used for wafer manufacturing and back-end operations. Global warming potentials are those defined in the 2007 Intergovernmental Panel on Climate Change (IPCC) Fourth Assessment Report [63]. Because wafer production dominates water consumption among the life-cycle stages, water use is reported as equivalent to the direct quantity utilized in wafer manufacturing. Land use is omitted from the inventory due to a lack of land use data for all life-cycle stages.

5.2.5 Electricity Generation Emissions Factors

The life-cycle GWP of electricity from coal, natural gas and large-scale hydroelectric and solar are taken from Pacca [84], while that of nuclear power is from Fthenakis [44]. EPA GWP emissions factors are used for geothermal and biomass electricity [38]. For non-greenhouse gases, only direct emissions from Santa Clara's

Table 5.2 Use phase power by technology node

Technology node (nm)	350	250	180	130	90	65	45	32
Year	1995	1998	1999	2001	2004	2007	2010	2013
Maximum rated Power (W)	14	23	25	61	84	104	146	158

electric utility (Pacific Gas and Electric) are included in the model. Average NO_x, SO_2, and mercury emissions for Pacific Gas and Electric's conventional fuel plants are taken from the EPA's eGrid database [37]. Water consumed in electricity generation is taken as the U.S. average of 1.76 l/kWh [68].

5.2.6 Use Phase Power

The average power requirements for logic chips are taken from the 2001–2010 International Semiconductor Manufacturing Roadmap reports [104] and, for 1995–1999, from manufacturers' specifications (Table 5.2)

The use phase power consumption of the chip is calculated assuming a product lifetime of 6,000 h (3 years, 8 h per day, 5 days per week, 50 weeks per year) consistent with the literature [25, 101] and a power supply efficiency of 70%. For the base case, a CPU activity rate of 17% is used, based on the SYSMark standard [89] and the average activities in a workday [60], to represent the applications and use patterns of a business user. For the upper bound, a data center case is represented with 33% activity rate, 24 hours per day operation (with 99% uptime), and a lifetime of 18 months. The lower bound characterizes a home use scenario, with 3 h of use per day, 250 days per year at 15% activity, for 5 years.

According to the ITRS industry average, chip power demand has more than doubled over the past decade. The steady increase in power requirements for logic chips is the main cause of rising energy-related life-cycle impacts, as shown in the Results section.

5.3 Results and Discussion

5.3.1 Global Warming Potential

Global warming emissions per die have risen at each successive technology node (Fig. 5.2). Use-phase electricity consumption generates the majority of life-cycle GWP impacts at all nodes, with an increasing share over time. Device power in the use phase contributes an average of 73% across all nodes, and 93% at 32 nm. If POU abatement is used for PFC-emitting wafer processes, as assumed for these results,

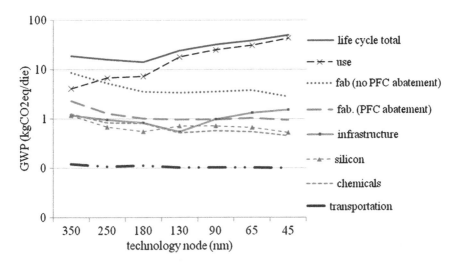

Fig. 5.2 Global warming potential per die by life-cycle stage, over 7 technology nodes

fabrication (including electricity and natural gas use, and direct emissions) produces on average 6% of life-cycle GWP over all years, and under 2% at the most recent generation. (Without PFC abatement, fab direct PFC emissions are 8 to 17 times higher and become the second largest single origin of GWP impacts after the use phase.) GWP from production of silicon and process chemicals account for about 10% of life-cycle GWP at the earliest 350 nm node, and their fractional contribution reduces over time, due largely to the increases in other life-cycle stages. Though the EIO LCI data for the GWP impacts associated with facility infrastructure (the fab building and equipment) are a high estimate for that activity, infrastructure clearly represents a substantial fraction of overall GWP impacts. (For further discussion of the sensitivity of life-cycle energy and GWP to production factors, see the chapter previous.)

5.3.2 Water Use

Life-cycle water consumption is dominated by electricity generation and the overall increase in water use is driven by climbing use-phase power as illustrated in Fig. 5.3. Water used in fabrication has fallen significantly per device over the period under study due to a number of changes in wafer processing. At the 130 nm node front-end-of-line (FEOL) photoresist removal steps switch from a wet sulfuric acid-hydrogen peroxide mixture (SPM) strip, to a dry plasma process which reduces the number of "wet" steps (in which the wafer is submerged in an UPW-based solution). As with transportation and fabrication emissions, the switch to 300 mm wafers at the

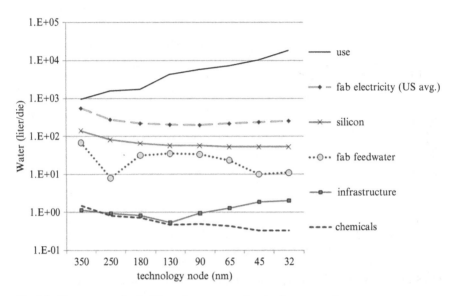

Fig. 5.3 Water use per die, by life-cycle stage, over 7 technology generations

130 nm node results in a reduction in water use but in this case the process change plays a greater role. Recycling of post-process UPW is not assumed in any of this study's fabs.

Land use is not considered in this study because consistent data concerning land utilization for processes throughout the life-cycle were not attainable, and the land occupied by the manufacturing sites would not be sufficiently representative of the entire life-cycle.

5.3.3 Photochemical Oxidant Formation, Acidification and Eutrophication Impacts via Air Emissions

Impacts from air emissions in the categories of photochemical oxidant formation (POF), acidification and eutrophication are primarily caused by use-phase electricity. Taking an average across all technology nodes, use phase electricity is accountable for over 80% of smog formation and eutrophication, and over 70% of acidification impacts via air. At the most recent, 32 nm technology node, use phase electricity causes over 95% of impacts for all of these categories. Over the generations, impacts related to use phase electricity have grown due to escalation of device power demand, which is the dominating factor defining the variation over time in POF, acidification, eutrophication and criteria health impacts per die. These impacts follow a common pattern, as shown in Figs. 5.4–5.7. Also contributing to the growth of these impacts is a steady rise in the size and complexity of fab

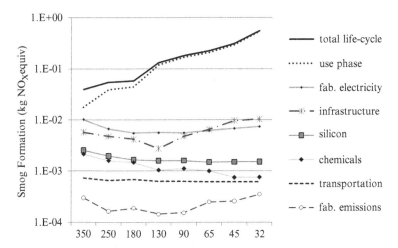

Fig. 5.4 Smog formation per die by technology node

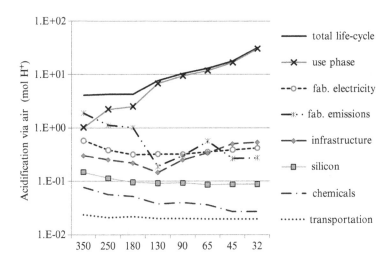

Fig. 5.5 Acidification per die by technology node

infrastructure. Steady increases in the complexity and length of the manufacturing process are the root cause of increases over the generations in impacts related to infrastructure, chemicals, and fabrication. Improvements to manufacturing efficiency – the switch from 200 to 300 mm wafers at the 130 nm node, and continuous reduction in the number of test and monitor wafers used over the first 6 generations – enable countervailing reductions in impacts in fab infrastructure, chemicals, silicon, transport, and fabrication impacts per die.

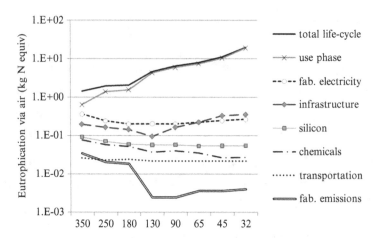

Fig. 5.6 Eutrophication via air emissions per die by technology node

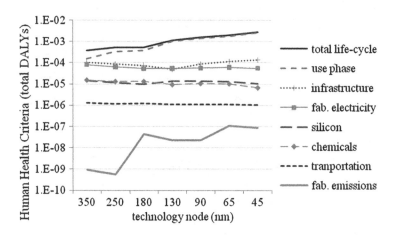

Fig. 5.7 EPA criteria human health impacts per die by technology node

5.3.4 Human Health Impacts

EPA criteria human health impacts are public health damages, measured in disability affected life-years (DALY), resulting from particulate matter, NO_x and SO_2 emitted during electricity production and transportation. Over the period of study, use phase electricity is the cause of an average of 77% of these health impacts. Life-cycle emissions from the fabrication building and equipment account for an average of 10% of criteria health impacts over the 18 year period (Fig. 5.7).

Non-cancer human health effects (such as developmental or neurological toxicity) are caused by the lead contained in chips produced before 1996 (350 nm

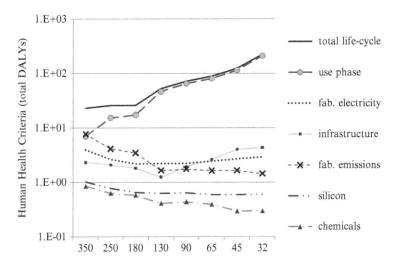

Fig. 5.8 Human non-cancer health impacts per die by technology node

through 90 nm nodes), electricity-related mercury emissions, lead emissions occurring throughout the upstream life-cycle of facility infrastructure as well as hydrochloric acid releases to water from wafer fabrication. At the most recent technology generation, 75% of non-cancer human health impacts are caused by use phase electricity, via mercury emissions, less than 1% are caused by fab direct emissions, 1% by fab electricity and 2% were due to facility infrastructure (Fig. 5.8). HCl vapor is abated with ≥99% efficiency in the fab scrubber and though liquid HCl is neutralized in the acid waste system with close to an equivalent efficacy, a small fraction may evade treatment (Fig. 5.8). Waterborne fab waste containing HCl flows to the municipal waste treatment system, where its actual impacts are lower than those predicted by direct release.

Human carcinogenicity results from the lead content of the chip disposed at end of life (EOL), as well as lead emitted in production of facility construction and manufacturing equipment and emissions from wafer fabrication. Among the fab emissions, formaldehyde and ammonia are the largest causes of these impacts. The v-shaped trend in impact magnitude over time reflects a decrease in fab-related emissions per die, due to yield improvements, combined with a steady increase in the size of the fab building and quantity of manufacturing equipment (Fig. 5.9).

Ecotoxicity impacts are largely due to airborne mercury emitted during electricity generation. EPA data for plants operated by Pacific Gas and Electric are estimated to emit 3.8 mg of mercury per MWh of electricity produced. The trend in ecotoxic impacts over time is therefore influenced by that of life-cycle electricity demand (Fig. 5.10). Fabrication emissions in the form of waterborne copper waste from die packaging and copper chemical mechanical polishing (CMP) account for less than 1% of ecotoxicity at all technology nodes.

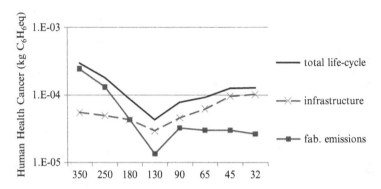

Fig. 5.9 Human carcinogenic impacts per die by technology node

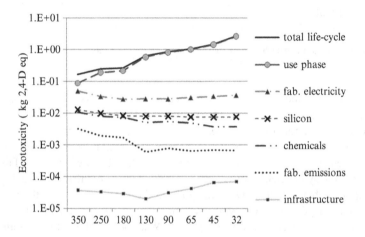

Fig. 5.10 Ecotoxicity per die by technology node

Evaluation of the eutrophication potential of waterborne emissions is compli-cated by the fact that modern fabs do not release any untreated liquid waste directly into the environment, but rather pass treated wastewater into the municipal sewer system. CMOS logic fabs use large quantities of both ammonia and sulfuric acid in wafer cleaning processes. Liquid ammonia may be recycled on site using membrane filtration or distillation, or sent to the acid waste neutralization (AWN) system. In this model, the latter is assumed and thus ammonium sulfate, which results from the neutralization of ammonia and sulfuric acid effluents, is among the liquid wastes produced in the highest volume by wafer fabrication. Ammonium sulfate may experience a number of possible reactions in the municipal wastewater treatment system, but unless the waste is de-nitrified in a bioreactor, it will likely be released to surface or coastal water as dilute ammonium. The worst case eutrophication impacts, assuming no downstream reaction or treatment of liquid waste, are presented in Fig. 5.11.

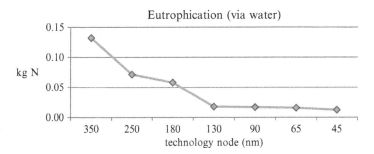

Fig. 5.11 Worst-case eutrophication impacts (via water) per die by technology node

5.4   Uncertainty

Abatement products have inherently high uncertainty because uncertainty in the mass of process exhaust is compounded by variation in abatement efficiency. (Abatement efficiency is typically defined as the molar ratio for a target species of abated output to system input.) The uncertainty associated with the output of abatement is thus always higher than input flows, but small mass flows of untreated materials have a particularly high uncertainty. For example, if the removal efficiency of a particular species falls between 97% and 99%, the uncertainty in the mass of untreated material will be ±50%. As the efficacy of treatment approaches 100% and the mass flow of untreated compound falls, uncertainty associated with that flow is magnified. Consequently, the uncertainty of impacts from fabrication emissions which are treated by facility abatement is much higher than for energy or water use, or PFC emissions, which have been measured post-abatement.

5.4.1   Sensitivity Analysis

Human health cancer impacts have the highest absolute sensitivity to lead emission factors for production of materials in the supply chain of fab equipment and building construction, and the highest relative sensitivity to abatement efficiencies. The impact category of water-bourne eutrophication, which results only from fab emissions, has the greatest sensitivity to fab abatement efficiencies for nitrogen compounds (mainly ammonia).

All other impact categories have the highest absolute sensitivity to use phase power demand and utilization, power supply efficiency, and to a lesser extent fabrication yields. A more detailed report of the quantitative sensitivity analysis of the model and findings for the sensitivity of life-cycle energy use and GWP to data uncertainty and model assumptions is presented in the previous chapter.

Table 5.3 VOCs lacking
POF impact metrics

TDMAT (tetrakis dimethylamino titanium)
TMS (tetramethyl silane)
TEOS (tetraethyl orthosilicate)
TMAH (tetramethylammonium hydroxide)
TDEAH (tetrakis-diethylamido-hafnium)
TDMAS (silicon tetrakis (dimethylamide))
DEA (diethamine)
PDMAT (petra-dimethyl amine tantalum)

5.4.2   Unaccounted Impacts

While less than 0.3% of life-cycle POF is due to wafer fabrication releases, 8 of the 17 VOCs used in semiconductor fabrication do not have impact factors, neither in TRACI nor among the EPA Reactivity Factors, and thus are unaccounted for in the model. While these releases amount to less than 0.02% of the total mass of VOCs throughout the life-cycle (contributing to about 40% of fabrication VOC releases by mass), the need for POF impact factors for the omitted chemicals (listed in the Supporting Information) are noted in Table 5.3.

Ecotoxicity, human health cancer and non-cancer impact factors also lack for a number of process chemicals (e.g., hexa-methyl disilizane, octamethyl-cyclotetrasiloxane, ruthenium compounds) which, while emitted in extremely small quantities, have undefined impact factors. Chemicals used during semiconductor manufacturing which lack human health and ecological impact factors are listed in Table 5.4. Though toxicological studies have been completed for most (if not all) of these materials, they lack impact factors.

Estimated components and proportions were used for CMP slurries, which have proprietary formulas. Knowledge of the exact composition of copper, oxide and tungsten CMP chemicals would enable more accurate LCA in the future.

Because many sources of process inventory data for chemicals describe only energy use, the impacts associated with chemicals production in this model only include primary energy use, water consumption, GWP, and no other emissions. This lack of information on environmental emissions from the production of chemicals is noted as an area of data scarcity.

Due to the lack of available data, values representative of chemical production of relatively low purity industrial products are used for chemicals of high purity. Data for the energy consumed in purification of semiconductor chemicals, particularly the elemental gases (N_2, Ar, He, O_2) and common process reactants such as silane, to semiconductor requirements (99.9999997%) would enable more accurate semiconductor LCA in the future.

Table 5.4 Chemicals lacking toxicity impact metrics

Al	N_2O
ArH	NF_3
Au	n-methyl-2-pyrrolidone
B_2H_6	NO
BCl_3	NO_2
BF_3	O_3
bis tertiary-butylamino silane	OMCTS (octamethyl-cyclo-tetrasiloxane)
BTA (benzotriazole)	p-cresol
Cl_2	PDMAT (petra-dimethyl amine tantalum)
CO	PGME (propylene glycol monomethyl ether)
COF_2	PGMEA (polypropylene glycol monomethyl ether acetate)
DCS (dichlorosilane)	PH_3
DEA (diethamine)	Pt
DMA (dimethyl amine)	$SiCl_4$
ethyl lactate	SiF_4
F_2	Ta
GeH_4	TDEAH (tetrakis diethylamido hafnium)
H_2O_2	TDMAS (tetrakis dimethylamide silicon)
H_2SO_4	TDMAT (tetrakis dimethylamino titanium)
HBr	TEOS (tetraethyl orthosilicate)
HCl (gas)	Ti
HF	TMAH (tetramethylammonium hydroxide)
HMDS (hexamethyldisilizane)	TMS (tetramethyl silane)
m-cresol	W
MMA (methyl methacrylate)	WF_6

5.5   Discussion and Conclusions

A complete and transparent LCA of semiconductor logic has hitherto been unavailable to LCA practitioners seeking to assess the impacts of electronic systems and devices. This lack of data has limited the analysis of information and communication technology (ICT) as a tool in energy efficiency and GWP goals. The efficacy of a particular application of ICT in reducing net energy use or GWP can be more definitively evaluated using the values for the environmental and human health impacts of semiconductor logic presented here. Many generations of CMOS logic are evaluated in this study, with earlier generations representing logic currently used in lower performance applications such as embedded logic in appliances and the later generations, computers and servers. The lack of life cycle inventory data for high purity chemicals and environmental impact factors for exotic or specialty chemicals continues to be a difficulty in semiconductor LCA, however. Listed in the appendix are process emission which may be of environmental concern but lack impact factors, as well as chemicals for which, in this study, a generic energy intensity value is used in place of specific inventory data.

By viewing the LCA results over time we can see trends in impacts per die. As use-phase power consumption and the complexity of fabrication have escalated so have electricity- and infrastructure-related emissions. Reductions in fabrication emissions have been achieved at certain points with a few key process changes, as well as the limitation of test and monitor wafer runs which results in higher line yield (the ratio of finished wafers to processed wafers). Average line yield increased from 68% to 88% in the period under study, but as this metric approaches its practical limit, its benefits taper. As device complexity lengthens the process flow, fabrication impacts per die can be expected to rise in the future unless process technologies and fab operations can continue to adapt to meet emissions targets.

Emissions due to electricity consumed in the use phase dominate most impact categories, particularly in the more recent technology generations. At the latest 32 nm technology node 96% of smog formation, 96% of acidification, 96% of eutrophication via air, 99% of ecotoxicity as well as 88% of human non-cancer health effects and 88% of EPA criteria health impacts are due to use phase electricity. Limiting use-phase power consumption, through technical or operational means, is the most effective way to limit the life-cycle impacts of digital logic as we go forward.

Chapter 6
Life-Cycle Assessment of Flash Memory[1]

6.1 Introduction

Solid state drives (SSD) show the potential for environmental benefits over magnetic data storage due to their lower power consumption. In this study, a complete life-cycle assessment (LCA) of flash memory over five technology generations (150 nm, 120 nm, 90 nm, 65 nm, and 45 nm) is presented to investigate this idea. The inventory of materials and energy used in NAND flash manufacturing is based on process data, while the impacts associated with chemicals, equipment, fab construction, electricity, and water are determined using economic input-output life-cycle analysis (EIO-LCA) and hybrid LCA. Sensitivity analysis shows that the most influential factors which can reduce the environmental impact of flash memory are perfluorinated compound (PFC) abatement in wafer fabrication and electricity use in manufacturing. A comparison between the complete life-cycle of 96 GB of flash memory and the use and transportation stages of a 2.5" hard disk drive (HDD) shows that the flash memory consumes less primary energy and water and results in lower global warming potential (GWP), ecotoxicity and smog formation than the magnetic HDD, though the relative benefit of flash-based SSD in other impact areas cannot be determined without a complete LCA of HDD.

Flash memory is one of the fastest growing semiconductor product types and is becoming competitive with magnetic hard disk drives (HDD) as computer storage. While solid state drives (SSD) are assumed to have a lower environmental impact than HDD because they require less power during operation, the life-cycle environmental impacts of Flash-based drives have not yet been studied. While SSD have low power consumption, their manufacture is highly energy and resource intensive as compared with most consumer goods. In this analysis, we present the life-cycle environmental impacts of NAND flash and compare the energy-related environmental impacts of SSD with those of HDD storage.

---

[1]Portions of this chapter are reprinted with permission from S. Boyd et al., *Life-Cycle Assessment of NAND Flash Memory*, IEEE Transactions on Semiconductor Manufacturing, February 2011. Copyright 2011 IEEE.

S.B. Boyd, *Life-Cycle Assessment of Semiconductors*, DOI 10.1007/978-1-4419-9988-7_6,   87
© Springer Science+Business Media, LLC 2012

Flash memory was developed from a combination of erasable, programmable read-only memory (EPROM) and electronically-erasable, programmable ROM (EEPROM) technologies in the mid-1980s and became widely produced for consumers in the mid-1990s. Because flash memory can store and access data with no moving parts, unlike magnetic storage, it has been applied to a variety of memory applications in consumer electronics and is widely used in digital music players and small-capacity, portable data storage. As a result, flash EPROM has been among the fastest growing types of semiconductor products in recent years [112, 113]. NAND and NOR flash are composed at the lowest level of transistors which implement logical NAND and NOR operations, respectively, with NAND being the denser but slightly slower design option. When the density of flash storage capacity recently reached 4 and 8 GB per cm^2 chip area, it became possible to package flash into products which could replace traditional hard disk drives (HDD). Flash-based solid-state drives (SSD) which are initially being introduced in laptops may also become competitive in desktop and data center applications, if scaling and cost challenges are overcome. Because flash memory is a fast growing semiconductor product segment which has the potential to expand further if SSD become more common in computer storage, the life-cycle environmental impacts are of particular interest.

6.2 Methodology

This study presents a life-cycle assessment (LCA) of five generations (150 nm, 120 nm, 90 nm, 65 nm, and 45 nm) of flash memory with wafer fabrication in Santa Clara, California, using chemicals, equipment and construction materials produced in the U.S. The mass of process chemicals consumed and emitted in each process step have been determined using in-line measurement, while equipment utility demands such as power, cooling water and utility nitrogen are based on both equipment measurements and specifications. Fab utility system capacities and resource demands are modeled using data from Sematech [104] and reflect industry-standard efficiency improvements over the 9 year period under study [15a]. Direct emissions from electricity generation are specific to California, based on data from the EPA [37] and primary energy use in electricity generation is taken from International Energy Agency data as 12 MJ/kWh, an average for the U.S. [99]. Life-cycle impacts due to water supply and product transportation are based on previous hybrid LCA studies [40, 118]. Chemical production and fab infrastructure (facility construction and equipment) are accounted for in this analysis using energy consumption and emissions determined using economic input-output LCA (EIO-LCA) [18]. Water consumed in the generation of electricity used in manufacturing, chemicals and fab infrastructure production and the product use phase is determined using a U.S. average of 1.76 l/kWh from a previous study [68]. IPCC global warming potential (GWP) impact factors are used for per-fluorinated compounds [63] and all other environmental and human health effects of emissions are evaluated using TRACI mid-point impact metrics, which are specific to the U.S. and California [81]. At end of life, it is assumed that there is no recoverable value from a discarded flash chip and the only significant impacts associated with disposal are the release of lead. The LCA methodology used in this study has been explained in further detail [15a] and inventory data reported on a per-process basis [71] in previous articles.

The functional unit of this study is one GB memory over a typical lifetime of 100,000 cycles. For all generations, wafer size is 300 mm and packaging is TSOP. Assumed line and wafer yields and a summary of the process technologies at each node are given in the appendix. All process flows and device memory capacities represent single-level cells (SLC, aka single-bit cells). Multilevel cells (MLC), which have become more widely produced in recent years, allow a doubling of bits per cell (or quadrupling in the case of 4xMLC). Because MLC can be manufactured without a significant increase in the number of steps in the manufacturing process flow versus SLC, MLC have roughly half of the environmental impacts as SLC per GB capacity. However, because MLC have shorter lifetimes and slower programming speeds than SLC and thus do not represent equivalent functionality, SLC are used throughout the study for consistency.

6.3 Results

Because use phase power per bit has been reducing or constant and the number of process steps required in wafer production has not increased considerably over these five flash technology nodes, the environmental impact of flash memory per chip has remained relatively flat over the past decade. Over the same period of time, device scaling as well as system-level enhancements of flash technology have allowed almost 16 times more memory capacity per device area. The combination of these trends results in a dramatic decrease in environmental impacts per unit of memory capacity for NAND Flash. An example of the results of these paired trends, primary energy consumption per gigabyte (GB) memory capacity by life-cycle stage is shown in Fig. 6.1. It should be noted, however, that despite the reductions in

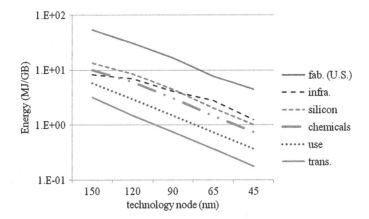

Fig. 6.1 Primary energy consumption per operating system equivalent memory capacity (MJ/GB), over five technology nodes

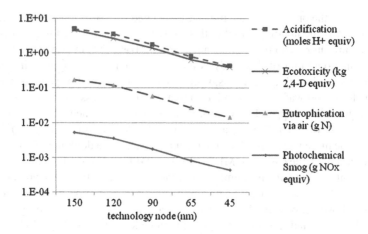

Fig. 6.2 Environmental impacts due to air emissions per GB, over five technology nodes

impacts per unit memory capacity, the environmental and human health impacts caused by flash memory as an industry, or flash memory product worldwide, is on the rise, due to the even more rapid expansion of the production and use of these products.

Flash scaling, for SLC, does not entail additional interconnect layers. For this, among other, reasons, the number of steps in the generic NAND process flow has not increased as rapidly as in the case of other semiconductor product types, particularly CMOS logic [15a]. Because the process flow has not expanded dramatically, direct emissions from wafer fabrication have not increased markedly over the period under study and, correspondingly, per-wafer impacts associated with the production of process chemicals have been relatively flat. In Fig. 6.2 the trends over the five technology nodes in ecotoxicity, acidification, eutrophication and smog formation are shown illustrating how minimal increases in per-wafer impacts result in notable reductions per GB. Ecotoxic impacts are due almost entirely to mercury emissions from electricity generation, with over 99% of life-cycle ecotoxicity coming from electricity generation and the remaining less than 1% due to formaldehyde emitted during wafer fabrication. About 50% of ecotoxic impacts are due to electricity used during manufacturing, a share which is also exemplified by the relative primary energy demand of manufacturing as shown in Fig. 6.1. Acidification impacts are caused by life-cycle emissions of oxides of nitrogen (NO_X) caused by fab infrastructure (constituting between 62% and 72% of life-cycle acidification over the five technology nodes), NO_X and SO_2 from transportation (17–25% of the total) and electricity generation (7–11%), and HF emissions from fabrication (1–10%). Eutrophication is attributable to NO_X emissions related to infrastructure (composing between 55% and 65% of these impacts over the five generations), transport (19–24%) and electricity generation (16–19%), with a small fraction (<2%) occurring as a result of fab gaseous emissions of NO_X and ammonia. The largest share (53–62%) of smog formation is caused by NO_X and CO emissions

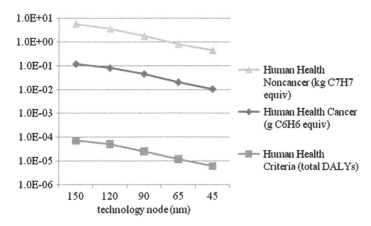

Fig. 6.3 Human health impacts due to air emissions per GB, over five technology nodes

produced due to fab infrastructure production, followed by NO_X and CO from transportation (17–23%) and electricity (15–18%). The remaining smog-forming impacts (4–7% of the life-cycle total) result from emissions (post-abatement) of isopropyl alcohol (IPA), CO, NO_X, ethyl lactate and other volatile organics from the fab.

Human health related impacts per wafer and device have shown the same stability over the past decade. Figure 6.3 shows human health impacts per GB over the five technology nodes. Non-cancer human health impacts (including developmental, reproductive and neurological toxicity) are primarily attributable to HF and other fluorine compounds, CO and dimethyl amine emitted, post-abatement, from wafer fabrication. Manufacturing represents between 66% and 72% of these non-cancer health impacts, with the remainder coming from infrastructure-related lead emissions (22–28%) and mercury released during electricity generation (6–7%). Carcinogenic human health effects principally result from manufacturing emissions of formaldehyde, which represent 72–75% of these impacts, while lead emissions resulting from fab infrastructure cause the remaining fraction. EPA criteria human health impacts, the most common public health threats resulting from smog, particulate matter (PM), lead, NO_2, SO_2 and CO, result from (in descending order) PM, SO_2 and NO_X emitted in throughout the supply chain in production of the manufacturing facility and equipment, which compose 68–75% of the life-cycle totals in this category over the period under study. SO_2 and NO_2 from electricity (19–23%) and transport (6–9%) also contribute to these human health effects.

Per-fluorinated compounds (PFCs) are an important group of emissions from semiconductor manufacturing due to their high infrared absorption, long lifetimes and consequential global impact. The World Semiconductor Council (WSC), which includes the semiconductor industry associations of Japan, Europe, Korea, Taiwan and the United States, has committed to PFC emissions reductions of 10% from 1995 or 1999 baseline levels by the end of 2010. However, in China, Singapore

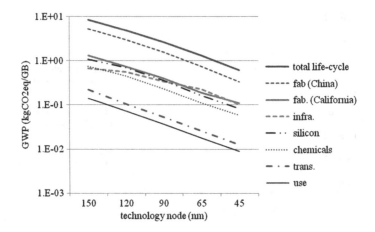

Fig. 6.4 GWP per GB memory capacity, by life-cycle stage, over five technology nodes

and Malaysia the semiconductor industry consortia have not made a commitment to control PFC emissions and in 2008, about 20% of semiconductor production capacity was held in these countries [9]. In Fig. 6.4, GWP impacts are shown by life-cycle stage with two scenarios illustrated, one in the U.S., where PFC abatement is necessary to meet the WSC goal, and the other in China, where there is no such resolution and PFCs are not abated.

In the U.S. example, direct emissions from wafer fabrication (CO_2, N_2O, methane and PFCs) cause less than 2% of life-cycle GWP, because PFCs are broken down using point-of-use (POU) abatement. The largest contributing cause of GWP is the electricity used in wafer fabrication and chip assembly, followed closely by silicon production, chemicals and fab infrastructure. The relative contribution of each of these life-cycle stages is shown in Fig. 6.4. If wafer fabrication is performed without PFC abatement, fab direct emissions constitute the largest fraction of GWP among all life-cycle stages and the total life-cycle GWP impacts of Flash memory increase by 24% to 30%, as demonstrated by the curve for fabrication and total life-cycle GWP for the China fab scenario in Fig. 6.4.

Water consumption is dominated by electricity generation, as shown in Fig. 6.5. At all technology nodes, water consumed in manufacturing represents less than 13% of life-cycle totals. (The fractional contributions of each life-cycle stage to total water consumption differ from those for primary energy use because not all energy use represents electricity.)

6.4 Discussion: Solid State Drives vs. Hard Disk Drives

Although a LCA of magnetic storage has not been reported in the literature, we find that considering only use-phase and transportation data for a hard drive, these HDD

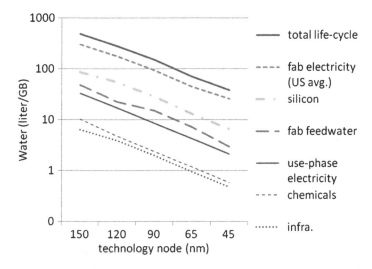

Fig. 6.5 Water consumption per GB memory capacity, by life-cycle stage, over the five nodes

impacts outweigh those of the full life-cycle of flash in most impact categories. A laptop-sized drive composed of 96 GB of 45 nm SLC flash (12×8 GB chips) is used for comparison with an equivalent capacity HDD. Each flash chip is assumed to have a mean time to failure of 100,000 erase cycles which, with wear leveling, allows a lifetime of at least 3–5 years. The magnetic HDD is assumed to have a read/write power of 4.7 W and an idle power of 3.6 W, based on an average of several manufacturers' specifications. With 20% active, 80% idle operation over a 4-year lifespan of 25,400 h, the HDD would consume 93.3 kWh of electricity. (Although lifetimes of HDD are variable, a life-span of 4 years in not uncommon and is chosen to match the functional unit.) The HDD use phase alone would result in 27 kg CO_2eq. in GWP, which is two orders of magnitude greater than the GWP impacts resulting from the full life-cycle of 96 GB flash. Each category of environmental and human health impacts are shown for the flash memory and limited inventory for HDD in Table 6.1.

6.4.1   Uncertainty

The environmental impact data with the greatest uncertainty range in the model are the emissions associated with fab construction and equipment production and the primary energy consumed in chemicals manufacturing. Due to the abstraction inherent in economic input-output modeling, EIO-LCA entails temporal and geographical uncertainty, as well as impact misallocation arising from generalization over each economic sector. The impacts associated with fab infrastructure and

Table 6.1 Full life-cycle of 96 GB flash memory vs. HDD transportation and use

		Complete life-cycle 96 GB flash	Use phase and transport of HDD
Primary energy consumption	MJ	724	1120
Water consumption	liters	3688	7173
GWP	kg CO_2 eq	0.54	27
Smog	kg NO_X eq	0.042	37
Acidification	moles H^+ eq	42	37
Ecotoxicity	g 2,4-D eq	38	86
Eutrophication	g N	1.4	1.3
Human health criteria	DALYs	5.8×10^{-4}	4.0×10^{-4}
Human health Noncancer	kg C_7H_7 eq	43	6.9
Human health cancer	kg C_6H_6 eq	8.6×10^{-5}	N/A

chemicals therefore have relatively high uncertainties, which are accounted for in the tabulated results (Appendix B.3). Fabrication emissions, because they are all post-abatement mass flows, have a high uncertainty that results from variation in the effective rate of facility abatement systems. An abatement system which operates at 99% efficiency with variation of ±1% produces mass flows of abatement products with a magnified uncertainty range of ±100%.

The device performance data with the greatest uncertainty in this study are the lifetimes assumed for the HDD and flash memory. Though a peer-reviewed empirical study of flash memory durability is not available, a 4-year life span for SLC flash is conservative [23]. While a percentage of NAND flash bits fail over the life of the chip, data checking algorithms compensate for lost bits and catastrophic breakdown of a flash device is rare (in contrast to HDD). The performance of a flash drive will nevertheless diminish over time, and thus the lifetime of a SSD is an inherently fuzzy value. The MTBF for the HDD in this analysis is chosen to match that of the flash memory and though a 4 year lifetime is supported by a previous large-population HDD reliability study [90], there is a wide uncertainty range associated with this value.

6.4.2 Sensitivity Analysis

By comparing the results for fabrication with and without PFC abatement, it is apparent that the most crucial decision affecting the life-cycle GWP of flash is the presence of PFC abatement in the fab. To determine the importance of other variables in the model, we use sensitivity analysis, testing the change in impact values with alterations in model parameter values. Sensitivity analysis shows that, because the largest fractions of environmental impacts ultimately result from emissions and resource consumption due to electricity generation, the emission factors for electricity have the greatest influence over the most impacts categories. Emissions from electricity generation cause the largest fraction of impacts in the categories of primary energy consumption, water consumption, GWP and ecotoxicity, and contribute a significant fraction to smog formation, eutrophication, acidification, and EPA criteria human health impacts. Impacts attributed to infrastructure and chemicals production are also ultimately caused by electricity used in the supply chain for these products. The energy sources and technologies used to generate electricity used in manufacturing and in the use phase, as well as in the supply chain of chemicals, equipment and fab construction materials, are the most critical factors which decide the magnitude of environmental and human health impacts.

The high uncertainties in the masses of emissions, as described in the previous section, have a significant influence on the certainty of the final life-cycle impact values, as reflected in the tabulated results in the appendix.

6.5 Conclusions

The results of this LCA also show that the largest shares of NAND flash life-cycle environmental impacts come from electricity generation and fab infrastructure production. Because the largest fraction of electricity is used in the manufacturing stage, it is most important to source fab electricity from low-impact sources. By siting a fab on an electrical grid supplied with a high percentage of renewable energy sources, or by supplementing grid-supplied electricity with on-site renewable generation, a flash manufacturer can most effectively reduce the life-cycle environmental impacts of its products. The second largest contributor to environmental and human health impacts overall is fab infrastructure production, which results in the largest proportion of impacts in the categories of smog formation, acidification, eutrophication and EPA criteria human health effects. Although all of the upstream activities associated with fab construction and equipment supply are difficult to control, minimizing the impacts associated with fab construction should also be a concern, due to the high resource and emissions intensity of construction activities and materials. The results of this model also show that, although overall human health impacts are modest, the largest fractions of human cancer and non-cancer health effects (besides EPA criteria impacts) occur as a result of direct fab emissions. Effective abatement and monitoring of fab emissions is essential to minimizing human health risks. Comparison between flash from facilities with and without PFC controls shows that without PFC abatement, PFC emissions cause the largest fraction of GWP impacts throughout the life-cycle. Abating PFCs is therefore the most important step towards reducing the global warming impact of flash memory.

This study validates the common assumption that flash-based SSD have a lower environmental impact versus HDD due to their significantly lower use-phase power consumption. Although the production of flash memory is highly energy and resource intensive, in the areas of primary energy and water consumption,

GWP, smog formation and ecotoxicity, they are lower than those produced in the transportation and use phases of a typical HDD. The relative impacts of SSD versus HDD in other impact categories can be determined with an environmental impact study for HDD of an equivalent scope to the study presented here, such that it includes manufacturing and tributary activities (e.g., materials production).

Chapter 7
Life-Cycle Assessment of Dynamic Random Access Memory

7.1 Introduction

Dynamic random access memory (DRAM) is the most common type of volatile memory and is a component of all laptop and desktop computers. In this study, life-cycle impacts of DRAM are determined for 250, 180, 130, 90, 70, and 57 nm technology nodes, representing DRAM manufactured in large scale production from 1997 through 2008. Primary energy and water consumption, as well as global warming potential, acidification, eutrophication, ground-level ozone (smog) formation, potential human cancer and non-cancer health effects, and ecotoxicity are evaluated. The life-cycle inventory is a hybrid model, using process data for wafer fabrication and die packaging, electricity production and some chemicals. Hybrid LCA data from previous studies are used for transportation and impacts associated with the water supply infrastructure. Economic input-output LCA (EIO-LCA) data from the Carnegie Mellon database [18] are used for the fabrication facility and manufacturing equipment, as well as some chemicals for which process data are unavailable.

Results are presented using, as the functional unit, the memory requirements to run an operating system (OS). As discussed in Chap. 4, the choice of functional unit has a great influence over how impact trends appear over time. While the environmental and human health impacts per gigabyte (GB) of DRAM have decreased over the time period of this study, those associated with the memory required to run an average desktop computer have not. The manufacturing inventory and life-cycle impact results on a per-GB basis, which are more readily usable by LCA practitioners, are presented in the appendix. Results in this chapter are presented per OS to investigate the influence of software memory efficiency on the life-cycle impacts of computing.

S.B. Boyd, *Life-Cycle Assessment of Semiconductors*, DOI 10.1007/978-1-4419-9988-7_7,   97
© Springer Science+Business Media, LLC 2012

7.2 Methodology

The life cycle inventory methodology follows that of the previous chapter, with the process flow for DRAM replacing that of flash memory. For each technology node, the fab facility system configurations for the corresponding year are used. Details on fab facility efficiency improvements are detailed in Chap. 3 (see Table 4.13. For each technology node, the minimum memory capacity required to run the latest version of the Windows OS, in the year corresponding to that node, is used. The memory requirements for the OS used at each technology node are given in Table 7.1. GWP intensity of electricity is specific to California (300 gCO_2eq/kWh), primary energy consumption in electricity generation is 12 MJ/kWh, an average for the US [99] and water consumption is 1.78 l/kWh, based on the most recent available (1995) USGS data [68].

Impact assessment also follows the methodologies of the previous chapters. TRACI mid-point impact metrics, which are specific to the U.S. and California [81], are used for smog formation, acidification, eutrophication, ecotoxicity and human health impact factors of emissions. IPCC global warming potential (GWP) impact factors are used for per-fluorinated compounds [63].

Chip sizes and memory capacity are based on International Technology Roadmap for Semiconductor standards for chip size at production and memory density [104] (Table 7.2). Chip power demand is determined using a manufacturer's data sheets [78], as listed in Table 7.2. The lifetime of the chip is assumed as 3 years, with 6,000 h of total use, consistent with the literature [25]. Power supply efficiency is accounted for in use phase energy consumption and is assumed to be 75% at all nodes.

Table 7.1 Year and OS memory requirements for each technology node

Year	Technology node (nm)	Windows RAM requirements (MB)
1997	250	4
1999	180	16
2001	130	32
2004	90	64
2006	70	64
2008	57	512

Table 7.2 Chip size and power demand for each technology node

Technology node (nm)	250	180	130	90	70	57
Memory capacity (GB)	0.128	0.256	0.512	1	2	2
Chip size (mm²)	128	176	127	93	110	74
Power demand (W)	0.35	0.38	0.47	0.50	0.53	0.53

7.3 Results

7.3.1 Resource Consumption

Primary energy consumption per OS memory capacity by life-cycle stage is shown
in Fig. 7.1. In recent years, the use phase is the largest contributor to life-cycle
energy consumption despite the low operating power of DRAM (less than 1 W).
Use phase energy consumption has increased by about 50% over the period of
study, while energy used in the wafer fabrication stage, per die, has decreased due
to smaller die sizes [104]. Energy used in supplying water to the fab is so small that
it falls well below the other contributions shown in Fig. 7.1, and is excluded to avoid
distortion.

Water consumption is dominated by electricity generation, as shown in Fig. 7.2.
The dominant consumers of water in the life-cycle of DRAM are fabrication
and use phase electricity. At all technology nodes, water consumed directly in
manufacturing (fab feedwater), e.g. to produce ultra-pure water, represents less than
8% of life-cycle totals.

7.3.2 Environmental Impacts

Members of the World Semiconductor Council (WSC), the semiconductor industry
associations of Japan, Europe, Korea, Taiwan and the United States, agreed in
1999 to limit PFC emissions using targets according to Kyoto-like baseline years.
However, the semiconductor industry associations in Singapore, Taiwan, Malaysia

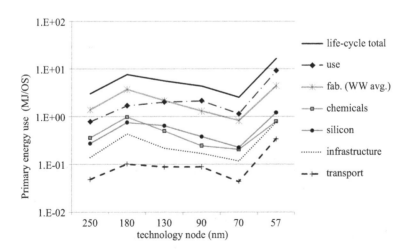

Fig. 7.1 Primary energy consumption per OS memory capacity (MJ/OS), over five technology
nodes

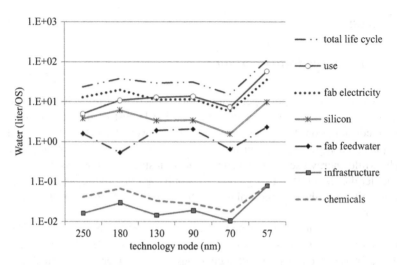

Fig. 7.2 Water consumption per OS memory capacity, by life-cycle stage, over five technology nodes

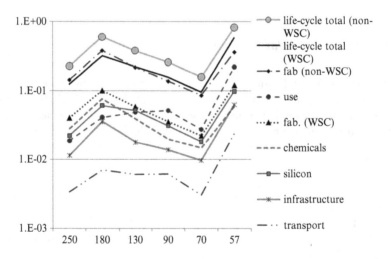

Fig. 7.3 GWP per OS memory capacity, by life-cycle stage, over five technology nodes

and China have not joined in the WSC agreement. To show the effect of the lack of PFC abatement in wafer fabrication, both WSC and non-WSC scenarios are presented for GWP impacts.

PFC abatement is a very important determining factor in life-cycle GWP as illustrated in Fig. 7.3. In the non-WSC case, PFC emissions are the greatest contributor to life-cycle GWP. Without PFC abatement, in the non-WSC scenario, life-cycle GWP jumps by between 42% and 88%.

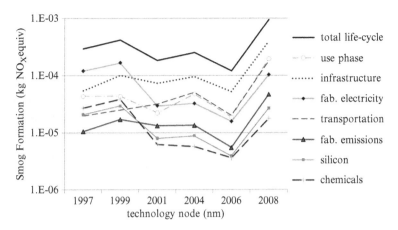

Fig. 7.4 Smog formation per OS by technology node

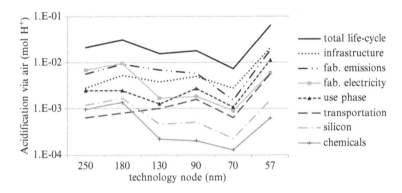

Fig. 7.5 Acidification per OS by technology node

The largest contributors to smog formation are fab infrastructure (building construction and semiconductor manufacturing equipment) production, use phase electricity, and transportation. The ground-level smog from fab infrastructure is ultimately due in large part to NO_X emissions from truck transportation and power used in the supply chain. Less than 8% of smog formation is attributable to direct volatile organic compound (VOC) emissions from wafer production (Fig. 7.4).

Infrastructure construction, fab direct emissions, fab electricity and use phase electricity are the largest factors in acidification impacts. The largest causes of acidification within fab infrastructure production are NO_X and SO_2 emitted from truck transportation, power generation and cement and aluminum manufacturing. HF emissions from wafer fabrication amount to, on average, about one third of life-cycle acidification for DRAM (Fig. 7.5).

The largest contributors to eutrophication are fab infrastructure, use phase electricity, transportation and fab electricity. The largest contributors to eutrophication

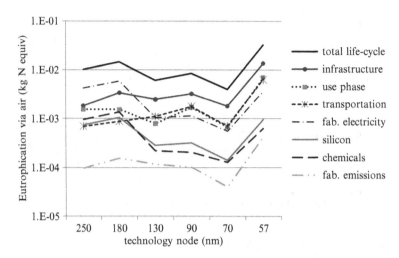

Fig. 7.6 Eutrophication via air emissions per OS by technology node

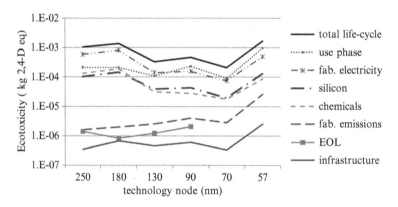

Fig. 7.7 Ecotoxicity per OS by technology node

occurring in the production of the fab facility infrastructure are NO_X emitted by truck and rail transportation, and in cement manufacturing. Only a very small fraction of eutrophication (about 1%) occurs as a result of gaseous emissions of NO_X and ammonia from wafer fabrication (Fig. 7.6).

Ecotoxicity is largely caused by mercury emitted in the generation of electricity used in production and the use phase, as shown in Fig. 7.7. Chemicals, silicon and infrastructure production also contribute to this impact category through electricity-related mercury emissions. Infrastructure production also contributes, to a small extent, via lead emissions which occur in mining and refining certain metals. Wafer fabrication contributes to ecotoxicity through trace emissions of formaldehyde, which amounts to less than 2% of life-cycle ecotoxicity.

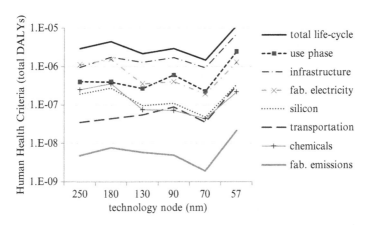

Fig. 7.8 Criteria human health impacts per OS by technology node

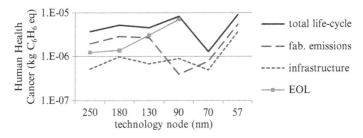

Fig. 7.9 Carcinogenicity per OS by technology node

7.3.3 Human Health Impacts

EPA criteria human health impacts, the most common public health threats resulting from smog, particulate matter (PM), lead, NO_2, SO_2 and CO. Life-cycle criteria health impacts for DRAM largely result from PM, SO_2 and NO_X emitted during production of the manufacturing facilities and equipment, and SO_2 and NO_X emitted due to electricity use in manufacturing and product use (Fig. 7.8).

Carcinogenic human health effects primarily result from manufacturing emissions of formaldehyde, followed by lead emitted in fab infrastructure production. Lead contained within the package of the DRAM chip, which is emitted into the environment at end of life (EOL) is a significant contributor in years up to 2006, when RoHS banned the presence of lead in electronics (Fig. 7.9).

Non-cancer human health impacts (including developmental, reproductive and neurological toxicity) are primarily attributable to HF and other fluorine compounds, CO and dimethyl amine emitted, post-abatement, from wafer fabrication. Mercury, released during generation electricity used in manufacturing and use, and lead emitted in the production of the fab facilities, are also substantial sources of

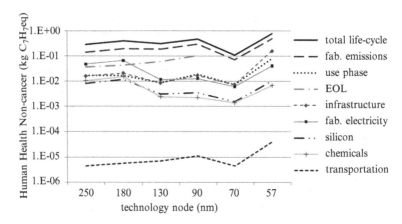

Fig. 7.10 Non-cancer human health impacts per OS by technology node

non-cancer health impacts. EOL lead emissions are also a significant contributor in years up to 2006. Health impacts from electricity (used in the use phase, wafer manufacturing and chemicals and silicon production) are principally due to mercury emissions from generating plants (Fig. 7.10).

7.4 Discussion

7.4.1 Comparison with Existing Work

Manufacturing inventory results from this study may be compared with a materials flow analysis of a DRAM chip by Williams [130]. The purpose of the Williams study was to bring attention to the energy and materials intensity of semiconductor production, which lies in contrast to the notion of de-materialization associated with the "information age." The subject of the paper is a 32 MB DRAM chip, which corresponds to 600 nm node DRAM, so a chip from the earliest technology node of this study, a 250 nm node chip with a capacity of 128 MB, is used for comparison. Given the earlier manufacturing technologies and simpler functional unit of the Williams paper, the results are expected to differ.

Williams investigates emissions as well as the energy and quantity of chemicals required to produce a DRAM chip using data from an anonymous industry member which he compares with several other data sources: a report from UNIDO, an unpublished study by an American industrial consortium, data from the Electronics Industry Association of Japan (EIAJ), and Toxics Release Inventory (TRI) data from the US government. Williams finds that the anonymous industry data is within an order of magnitude of the mass inventory data from all of these data sources except TRI, which is known to be incomplete. Williams stressed that the process inventory

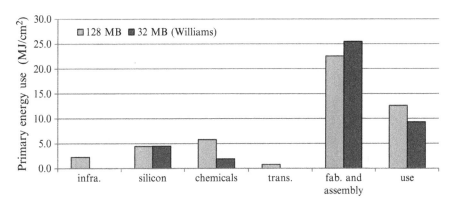

Fig. 7.11 Comparison with Williams: Energy use at each life-cycle stage

data he reported was not exact, but he cross-checks the mass inventory with several other industrial and government sources, it is worthwhile to use it for comparison, with an understanding of the difference in functional unit, and limitations due to data availability of that study. The Williams data is normalized to 1 cm² finished wafer area and adjusted to account for additional factors which are included in this analysis but not in his: line yield (which accounts for non-product wafers run in production) as well as lifetime of the chip (Williams assumes a 4,380 hour life) and power supply efficiency in the use phase.

Comparison of the energy and materials inventory results of the two studies shows that they differ but fall within an order of magnitude of one another. The divergence in results for energy use in fabrication is due to the fact that Williams uses an average of industry-level energy consumption for the semiconductor industries of Japan and the US to determine this value, which includes demand beyond the manufacturing facilities themselves, as well as data from an unpublished industrial consortium for 150-mm wafer production, which is less efficient than 200-mm wafer production. In the use phase, chips at the 250 nm node consume more power per unit area than chips at the 500 nm node, resulting in higher use phase energy expenditure of the 128 MB chip (Fig. 7.11).

While the totals for all chemicals and elemental gases used in production are similar between the two studies, the breakdown between elemental gases and other chemicals differ significantly. This disparity in materials inventory data is not clearly attributable to the difference in process technology between the two generations. Currently, and since the 180 and 130 nm nodes, certain wafer cleaning and thermal steps are performed in single-wafer systems which have higher per-wafer material demands. One of the most significant changes in material efficiency in semiconductor processing has been this switch, for certain more delicate steps, from batch process wet cleaning and thermal steps, to single-wafer cleaning and rapid thermal processing (RTP). In batch-based wet cleans, the cleaning solution can be more easily recycled, and, in furnace-based thermal processing, many wafers

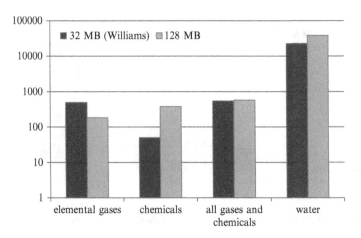

Fig. 7.12 Comparison with Williams: Mass of process chemicals and gases

would be annealed or oxidized at once, requiring a smaller quantity of process chemicals and gases per wafer. At the 250 nm node wet cleaning and thermal steps are performed by batch process, as at the 500 nm node, so demands for inputs to individual wet process or thermal steps should be similar between the two studies. At the 250 nm node, however, there are more layers of metallization as well as additional steps used for electrical isolation of the gate and likely more implant steps to control distribution of dopants in the channel area. Because the process technology of the DRAM in the Williams study is not described, it is difficult to say with certainty that process technology is not a major factor, but, because the process flow necessary to construct 128 MB DRAM entails more steps than that for 32 MB DRAM, the device complexity is the clearest cause of the discrepancy. Also, the data for process inputs in the Williams study are described as an estimation provided by an expert within industry, so the values for the mass inventory are approximate. Because the inventory results of this and the Williams study fall within an order of magnitude, the difference can be attributed, generally, to the simpler functional unit, as well as the approximate nature of the Williams data (Fig. 7.12).

Water use results compare as expected between the two studies, as the production of a later device technology will have higher water demands, due to the relatively larger number of wafer cleaning steps.

It is important to point out that one assumption stated by Williams should not be repeated in future semiconductor process inventory studies:

"Given that nearly all chemicals used in semiconductor processing do not end up in the final product, mass balance dictates that use and emissions of chemicals should be nearly identical."

Although this is true at the chamber level, and makes sense from a theoretical standpoint for the fab as a whole, the masses of input materials and emissions at the facility level differ in actuality due to chemical transformations which occur during processing and abatement with elements which may not be counted as "inputs" to the process (air, water and abatement materials). The largest mass fractions of semiconductor process waste are liquids used in wafer cleaning (HF, H_2SO_4, HCl, H_2O_2, NH_4OH). All of these liquids, with the exception of HF, are diluted and neutralized on site to a pH which is safe to drain into the public wastewater system. Because these wastes are non-toxic and can be released safely, they may not be counted as emissions by some studies. Also, in some localities, environmental permits for these releases are regulated by pH limits, rather than by the mass of waste flow, so there is no reason to measure or monitor the mass of these wastes. Fluorinated chemicals and HF liquids are converted into a matrix of calcium fluorite and flocculants, which is usually hauled off-site as non-hazardous waste. In this case, a record of the waste stream would be reported but the mass of materials in the waste product would dramatically over-estimate the mass of the initial process chemicals. One gram of HF used in process would be converted to almost seven grams of non-hazardous waste. While mass balance closes at the chamber level (minus chamber wall residues in the case of deposition steps), the same is not true at the facility level, and the mass of total waste from a fab cannot be used as a proxy for the mass of input process materials.

7.5 Conclusions

The overall trends in impacts reveal that software efficiency is an important factor in determining the life-cycle environmental impacts of computer memory. While impacts per GB memory capacity have fallen over the past decade, the life-cycle environmental and human health effects associated with the memory requirements of a typical operating system have risen.

At the most recent technology nodes, use phase energy consumption has been the largest contributor to energy and water consumption, ecotoxicity and, in the case of wafer production within the WSC, GWP impacts. Infrastructure production is the largest cause of smog formation, acidification, eutrophication, and human health criteria. Fab direct emissions are the largest source of carcinogenicity and non-cancer human health impacts. Results from this study compare with results from a previous manufacturing inventory for DRAM as anticipated, given the differences in data sources and functional unit.

Chapter 8
Semiconductor LCA: The Road Ahead

In this book, LCAs of CMOS logic, flash memory and DRAM are presented. Together, these LCA results enable a wide variety of electronic devices to be modeled, and thus allows comparison between a broad set of electronic devices or IT services with traditional products. The manufacturing inventory and life-cycle impact data provided in the appendices, which allow more accurate life cycle assessment of a wide variety of electronics, represent the most important contributions of this work. These LCA are transparent and include several sources of impacts which have not been included in the scope of previous published semiconductor LCA.

One of the major challenges to LCA of semiconductors is the rapid advancement of technology and requisite frequent changes to manufacturing processes. The use of process specific data is a particular strength of the LCA in this book, because using data at the equipment level rather than the fab level allows a more adaptable inventory model which has allowed the analysis of various devices. By applying the material and energy demands and chemical emissions for a set of processes to a device-specific process flow, many different vintages of a semiconductor device can be evaluated, including semiconductor products which have not yet reached full-scale production.

Among the results of the LCA in this book, the global warming impacts associated with PFC emissions from wafer manufacturing are particularly relevant to policy. As stated in Chap. 4, semiconductor manufacturing capacity is growing most rapidly in countries which lack government or industrial policy to control PFC emissions. The LCA results for CMOS logic, flash memory and DRAM show that emitting PFCs without abatement results in a dramatic increase in life-cycle GWP for these products. Increasing awareness of the link between products and the PFC emissions which occur during their production encourages commitment to quantitative PFC emission reduction goals.

An important conclusion concerning the functional unit in semiconductor LCA arose in the process of performing LCA for these products. Choice of functional unit is a not a straightforward decision for semiconductor products because, as software has increased its computational requirements in response to the greater

S.B. Boyd, *Life-Cycle Assessment of Semiconductors*, DOI 10.1007/978-1-4419-9988-7_8,   109
© Springer Science+Business Media, LLC 2012

supply of computing capacity, a given amount of computational power and memory
has provided diminishing functionality, by some measures. This effect, which was
described by Plepys as a rebound effect [91], has been explored further in this book
by charting the life-cycle impacts of the minimum random access memory (RAM)
requirements for a popular operating system (OS) over time. While the impacts per
GB of DRAM have fallen over the past decade, the impacts associated with the
amount of DRAM required to run the OS have not. An important finding of this
book is that software efficiency is a critical factor in the environmental impact of
computing.

8.1 Future Work

8.1.1 Life-Cycle Inventory and Impact Factors for Semiconductor Chemicals

The life-cycle stage for which data collection proves the most difficult is still
the production of high-purity and specialty chemicals used in wafer fabrication.
There continues to be a considerable need for life-cycle inventory data for the
energy used in and emissions from semiconductor chemicals production. While
LCI data for many common industrial chemicals have been defined and populate
LCA databases and software, equivalent information for chemicals which are
specific to semiconductor production are not currently available. Many specialty
chemicals used in wafer fabrication also do not have defined environmental or
human health impact factors. In order to improve the accuracy of semiconductor
LCA, energy consumption and emissions data for the production of semiconductor
grade (>99.9999997%) elemental gases (N_2, Ar, He, O_2). LCI data and impacts
factors for certain specialty chemicals which are used in high volume, particularly
CMP agents, would be most helpful.

For most chemicals used in these LCA, EIO-LCA data or a generic inventory
value for organic or inorganic chemicals was used. Where process data was available, it represents industrial grade rather than the high-purity form. The sensitivity
analysis of the CMOS logic LCI model provided in Chap. 4 shows that, even with an
uncertainty of ±50% or more for most materials, the uncertainty in chemicals LCA
data did not result in a significant increase in uncertainty in the final LCA results.
In the cases of DRAM and flash memory, which have much lower use phase power,
the uncertainty of chemicals LCA data results in a greater, though still manageable
range of uncertainty in the final results. These conclusions are made based on a
conservative (large) range of uncertainty for these data, but without knowing their
actual range of values, their influence on the final results cannot be known with
complete certainty.

The category of the process chemicals which currently have the highest consumption by mass in logic production, after elemental gases, but which lack LCI

or impact data, are CMP slurries. CMP slurries have proprietary formulas, so the components and proportions can only be estimated from basic process knowledge. Aggregate data for the energy and emissions of production, and impact factors for each type of CMP slurry, or the exact composition for generic copper, oxide and tungsten CMP chemicals would enable more accurate LCA in the future.

Following elemental gases and CMP chemicals, agents used in photolithography are among the chemicals used in highest volume in wafer processing. These materials are for the most part volatile organics, which have the potential to produce ground-level ozone and smog, though they may also have some health effects or toxicity. A list of the volatile organic chemicals which lack environmental impact and human health factors are given in Table 5.3.

There are several semiconductor chemicals which have undergone toxicity studies but lack environmental and human health impact factors. A list of the chemicals which are used in highest volume, and which show the potential to damage environmental or human health but lack impact factors are listed in Table 5.4.

8.1.2   LCA of Networked Thin Client Versus Local Desktop Computing

As more computing becomes dependent on the internet, some users are switching from larger, independent personal computers to thin clients, such as smart phones and "netbooks." These electronics, while having lower individual power consumption than a desktop PC require, constant interaction with servers over wireless internet, and in some cases also telecommunications networks. While the trend towards thin clients appears on the surface to be a positive environmental trend, due to their lower power consumption and smaller physical dimensions, the impacts associated with operation of network equipment and servers also need to be considered in the comparison.

8.1.3   Comparative LCA of IT Versus Traditional Products and Services

There are many applications for the data developed in this book to be used in LCA of electronics, which would be particularly useful for comparison with traditional products and services. Two good examples of these types of studies are a comparative life cycle analysis of reading news on a newspaper versus on a hand-held device [123] and a study of the life-cycle impacts of telecommuting versus commuting to work [69]. Other products and services that deserve analysis are the use of electronic books, online videos, internet encyclopedias and online banking in place of their traditional hard media equivalents.

8.1.4 Further Study of the "Rebound Effect" of Computational Power and Memory Capacity

As computational power and memory have become cheaper and more accessible, software has expanded to consume the processor capability and space afforded to it, and computer users have consumed more processing power and memory capacity. In this book, this effect is tested in the case of the memory capacity required for a standard operating system. This area of study can be expanded in both quantitative and qualitative directions. For example, the computational power for operating systems and common applications may be evaluated to illustrate trends in processor demand over time. Also, this topic can also be addressed at higher level of analysis. As technology advances and software applications consume more computational and memory capacity, users benefit from the depth and variety of tasks that software can perform. Some applications, given the additional computational power, have evolved beyond their original purposes. For example, internet browsing was originally used primarily for information and entertainment, and is now used for video communication, shopping, banking, interactive blogging and more. At the same time, some computer applications have continued to perform the same tasks (email communication, word processing, spreadsheet calculation), only using much more memory. Whether additional resources used to enhance existing applications and perform new tasks are being used effectively and efficiently is a complex question, but nevertheless an interesting area of deeper inquiry.

The manufacturing inventory and LCA impact data presented in this book enable more accurate and confident study of the potential environmental benefits of IT. My hope is that this research facilitates more informed and deliberate adoption of IT as a replacement for existing products and services, to allow continued technological progress, innovation and growth without damaging or endangering the environment or human health.

Appendix A
CMOS Logic: Manufacturing Inventory and Impact Data

A.1 CMOS Logic Manufacturing Inventory Data

Table A.1 Energy consumption per die by life-cycle stage

Energy (MJ/die) Year	1995	1998	1999	2001	2004	2007	2010	2013
Technology node (nm)	350	250	180	130	90	65	45	32
Total life-cycle (WW avg.)	286	329	336	720	976	1,200	1,668	2,963
Use	151	250	271	666	917	1,135	1,593	2,884
Fab (China)	100	50	40	37	36	40	43	47
Fab. (WW avg.)	86	43	34	32	31	34	37	40
Infra.	14	12	10	7	12	16	24	25
Silicon	15	13	10	8.9	8.9	8.3	8.3	8.3
Chemicals	18	10	9.1	5.8	6.2	5.5	4.1	4.1
Transportation	1.8	1.6	1.6	1.5	1.5	1.5	1.4	1.4

Note: Life-cycle totals have an accuracy of two significant digits, but are reported here unrounded

Table A.2 Water consumption per die by life-cycle stage

Year	1995	1998	1999	2001	2004	2007	2010	2013
Technology node (nm)	350	250	180	130	90	65	45	32
Use	969	1,602	1,736	4,264	5,872	7,270	10,206	18,469
Fab feedwater	68	8	32	36	34	24	10	11
Fab electricity (US avg.)	549	275	219	203	199	218	238	255
Silicon	140	81	66	57	57	53	53	53
Infrastructure	1.1	0.9	0.8	0.5	1.0	1.3	1.9	2.0
Chemicals	1.5	0.8	0.7	0.5	0.5	0.4	0.3	0.3
Life cycle water	1,729	1,968	2,054	4,561	6,164	7,566	10,510	18,791

Note: Life-cycle totals have an accuracy of two significant digits, but are reported here unrounded

S.B. Boyd, *Life-Cycle Assessment of Semiconductors*, DOI 10.1007/978-1-4419-9988-7,

Table A.3 GWP per die by life-cycle stage

GWP (kgCO2eq/die) Year	1995	1998	1999	2001	2004	2007	2010	2013
Technology node (nm)	**350**	**250**	**180**	**130**	**90**	**65**	**45**	**32**
Total life-cycle (50% abatement)	14	13	12	22	29	36	48	84
Use	4.1	6.8	7.4	18.1	24.9	30.8	43.2	78.2
Fab. (no PFC abatement)	9.5	5.0	3.3	3.1	3.0	3.6	3.0	3.3
Fab. (ideal PFC abatement)	2.5	1.3	1.0	0.9	0.9	1.0	1.1	1.2
Infra.	1.2	1.0	0.8	0.5	1.0	1.3	1.94	2.08
Silicon	1.2	1.0	0.8	0.7	0.7	0.7	0.7	0.7
Chemicals	1.5	0.9	0.8	0.5	0.5	0.5	0.5	0.5
Trans.	0.12	0.11	0.11	0.10	0.10	0.10	0.10	0.10

Table A.4 CMOS manufacturing inventory per wafer, 350 nm node

350 nm Inputs	g/wafer	350 nm Inputs	g/wafer
GWG		*Flammable*	
CF_4	4.74E−02	SiH_4	8.39E−03
CHF_3	8.69E−04	H_2	2.20E−01
C_2F_6	0.00E+00	DCS	0.00E+00
CH_4	4.36E−01	B_2H_6	0.00E+00
CO_2	0.00E+00	C_4F_8	0.00E+00
NF_3	2.02E−02	GeH_4	0.00E+00
C_4F_6	0.00E+00	C_2H_2	0.00E+00
N_2O	0.00E+00	C_2H_4	0.00E+00
Hazardous			
F_2	0.00E+00	*Other chemicals*	
HF (gas)	0.00E+00	O_2	1.20E−01
SiF_4	4.41E−05	Ar	6.56E−01
C_4F_8	3.24E−04	N_2	4.86E−01
C_2F_4	0.00E+00	CMP polishing solution	0.00E+00
COF_2	0.00E+00	H_2O_2	1.04E+01
HCl (gas)	0.00E+00	abrasive	0.00E+00
Cl_2	5.38E−03	silica slurry	0.00E+00
$SiCl_4$	0.00E+00	benzotriazole	0.00E+00
PH_3	2.45E−04	CMP surfactants	2.35E−01
BTBAS	1.05E−03	citric acid	1.31E−01
AsH_3	0.00E+00	tungsten CMP slurry	1.13E+01
BCl_3	6.25E−03	Al	3.74E−04
Pb	1.10E−05	As	0.00E+00
WF_6	7.99E−02	Cu	1.55E−04
HBr	4.33E−03	Pt	0.00E+00
Br_2	0.00E+00	Ti	2.61E−05
HMDS	1.36E−01	He	3.66E−04
OMCTS	0.00E+00	NH_3	1.11E−02

Table A.5 CMOS manufacturing inventory per wafer, 350 nm node

350 nm		350 nm	
Inputs	g/wafer	Inputs	g/wafer
Criteria		*Other chemicals*	
CO	0.00E+00	$CuSO_4$	2.04E−03
NO	0.00E+00	ArH	0.00E+00
NO_2	0.00E+00	100% HF liquid	2.06E−01
SO_2	0.00E+00	H_2SO_4	1.13E+02
N_2O	3.59E−02	HCl	2.31E+00
O_3	0.00E+00	NH_4OH	4.12E+00
		IPA	5.73E+00
VOC		oxide CMP slurry, chemicals	1.30E+01
DMA	0.00E+00	BF_3	0.00E+00
MMA	0.00E+00	Cr	0.00E+00
TDMAT	4.94E−04	Au	4.48E−05
TMS	0.00E+00	Sn	3.04E−06
TEOS	0.00E+00	Co	2.39E−05
formaldehyde	1.61E−02	Ni	2.01E−03
m-cresol	1.02E−02	polyimide laminate	2.03E−03
p-cresol	1.01E−02	laminate solvent	0.00E+00
PGMEA	1.30E−01	H_3PO_4	0.00E+00
n-methyl-2-pyrollidone	9.09E−03	$CuCl_2$	1.17E−06
PGME	1.71E−01	W	4.15E−07
ethyl lactate	1.64E−01	Ta	0.00E+00
TMAH	1.76E−02	Cu ECP additive X	5.93E−04
TDEAH	0.00E+00	Cu ECP Additive Y	2.94E−04
TDMAS	0.00E+00	$Na_2B_4O_7$	7.63E−03
DEA	0.00E+00		
PDMAT	0.00E+00		

Table A.6 CMOS manufacturing inventory per wafer, 250 nm node

250 nm		250 nm	
Inputs	g/wafer	Inputs	g/wafer
GWG		*Flammable*	
CF_4	5.51E−02	SiH_4	8.73E−03
CHF_3	9.43E−04	H_2	2.56E−01
C_2F_6	0.00E+00	DCS	0.00E+00
CH_4	4.75E−01	B_2H_6	0.00E+00
CO_2	0.00E+00	C_4F_8	0.00E+00
NF_3	2.10E−02	GeH_4	0.00E+00
C_4F_6	0.00E+00	C_2H_2	0.00E+00
N_2O	0.00E+00	C_2H_4	0.00E+00
Hazardous			
F_2	0.00E+00	*Other chemicals*	
HF (gas)	0.00E+00	O_2	1.21E−01
SiF_4	4.41E−05	Ar	7.37E−01
C_4F_8	3.24E−04	N_2	6.59E−01

(continued)

Table A.6 (continued)

250 nm		250 nm	
Inputs	g/wafer	Inputs	g/wafer
C_2F_4	0.00E+00	CMP polishing solution	0.00E+00
COF_2	0.00E+00	H_2O_2	1.14E+01
HCl (gas)	0.00E+00	abrasive	0.00E+00
Cl_2	6.12E−03	silica slurry	0.00E+00
$SiCl_4$	0.00E+00	benzotriazole	0.00E+00
PH_3	2.45E−04	CMP surfactants	2.82E−01
BTBAS	1.05E−03	citric acid	1.57E−01
AsH_3	0.00E+00	tungsten CMP slurry	1.36E+01
BCl_3	7.30E−03	Al	4.68E−04
Pb	1.10E−05	As	0.00E+00
WF_6	9.32E−02	Cu	2.67E−04
HBr	4.33E−03	Pt	0.00E+00
Br_2	0.00E+00	Ti	2.61E−05
HMDS	1.51E−01	He	7.32E−04
OMCTS	0.00E+00	NH_3	1.11E−02

Table A.7 CMOS manufacturing inventory per wafer, 250 nm node

250 nm		250 nm	
Inputs	g/wafer	Inputs	g/wafer
Criteria		*Other chemicals*	
CO	0.00E+00	$CuSO_4$	2.04E−03
NO	0.00E+00	ArH	0.00E+00
NO_2	0.00E+00	100% HF liquid	2.51E−01
SO_2	0.00E+00	H_2SO_4	1.21E+02
N_2O	4.09E−02	HCl	2.62E+00
O_3	0.00E+00	NH_4OH	4.75E+00
		IPA	6.05E+00
VOC		oxide CMP slurry, chemicals	1.52E+01
DMA	0.00E+00	BF_3	0.00E+00
MMA	0.00E+00	Cr	0.00E+00
TDMAT	9.88E−04	Au	7.69E−05
TMS	0.00E+00	Sn	3.04E−06
TEOS	0.00E+00	Co	4.10E−05
formaldehyde	1.79E−02	Ni	3.45E−03
m-cresol	1.14E−02	polyimide laminate	2.03E−03
p-cresol	1.13E−02	laminate solvent	0.00E+00
PGMEA	1.45E−01	H_3PO_4	0.00E+00
n-methyl-2-pyrollidone	9.09E−03	$CuCl_2$	1.17E−06
PGME	1.90E−01	W	4.15E−07
ethyl lactate	1.82E−01	Ta	0.00E+00
TMAH	1.96E−02	Cu ECP additive X	5.93E−04
TDEAH	0.00E+00	Cu ECP Additive Y	2.94E−04
TDMAS	0.00E+00	$Na_2B_4O_7$	8.48E−03
DEA	0.00E+00		
PDMAT	0.00E+00		

Table A.8 CMOS manufacturing inventory per wafer, 180 nm node

180 nm		180 nm	
Inputs	g/wafer	Inputs	g/wafer
GWG		*Flammable*	
CF_4	2.92E$-$03	SiH_4	9.31E$-$03
CHF_3	9.46E$-$04	H_2	2.73E$-$01
C_2F_6	5.94E$-$04	DCS	0.00E$+$00
CH_4	3.89E$-$01	B_2H_6	0.00E$+$00
CO_2	0.00E$+$00	C_4F_8	0.00E$+$00
NF_3	9.96E$-$02	GeH_4	0.00E$+$00
C_4F_6	0.00E$+$00	C_2H_2	0.00E$+$00
N_2O	0.00E$+$00	C_2H_4	0.00E$+$00
[2pt] *Hazardous*			
F_2	0.00E$+$00	*Other chemicals*	
HF (gas)	0.00E$+$00	O_2	8.24E$-$01
SiF_4	2.03E$-$03	Ar	2.67E$+$00
C_4F_8	3.01E$-$04	N_2	6.43E$-$01
C_2F_4	0.00E$+$00	CMP polishing solution	0.00E$+$00
COF_2	0.00E$+$00	H_2O_2	1.17E$+$01
HCl (gas)	0.00E$+$00	abrasive	0.00E$+$00
Cl_2	6.40E$-$03	silica slurry	0.00E$+$00
$SiCl_4$	0.00E$+$00	benzotriazole	0.00E$+$00
PH_3	2.28E$-$04	CMP surfactants	3.07E$-$01
BTBAS	9.75E$-$04	citric acid	1.70E$-$01
AsH_3	0.00E$+$00	tungsten CMP slurry	1.48E$+$01
BCl_3	7.77E$-$03	Al	5.23E$-$04
Pb	1.02E$-$05	As	0.00E$+$00
WF_6	9.92E$-$02	Cu	3.31E$-$04
HBr	4.03E$-$03	Pt	0.00E$+$00
Br_2	0.00E$+$00	Ti	2.43E$-$05
HMDS	1.54E$-$01	He	6.82E$-$04
OMCTS	0.00E$+$00	NH_3	1.04E$-$02

Table A.9 CMOS manufacturing inventory per wafer, 180 nm node

180 nm		180 nm	
Inputs	g/wafer	Inputs	g/wafer
Criteria		*Other chemicals*	
CO	0.00E$+$00	$CuSO_4$	1.90E$-$03
NO	0.00E$+$00	ArH	0.00E$+$00
NO_2	0.00E$+$00	100% HF liquid	2.52E$-$01
SO_2	0.00E$+$00	H_2SO_4	1.27E$+$02
N_2O	5.27E$-$03	HCl	2.58E$+$00
O_3	0.00E$+$00	NH_4OH	4.82E$+$00
		IPA	5.34E$+$00
VOC		oxide CMP slurry, chemicals	1.62E$+$01
DMA	0.00E$+$00	BF_3	0.00E$+$00
MMA	0.00E$+$00	Cr	0.00E$+$00
TDMAT	9.21E$-$04	Au	9.52E$-$05
TMS	0.00E$+$00	Sn	2.83E$-$06

(continued)

Table A.9 (continued)

180 nm		180 nm	
Inputs	g/wafer	Inputs	g/wafer
TEOS	0.00E+00	Co	5.08E−05
formaldehyde	1.84E−02	Ni	4.27E−03
m-cresol	1.16E−02	polyimide laminate	1.89E−03
p-cresol	1.15E−02	laminate solvent	0.00E+00
PGMEA	1.49E−01	H_3PO_4	0.00E+00
n-methyl-2-pyrollidone	8.47E−03	$CuCl_2$	1.09E−06
PGME	1.95E−01	W	3.86E−07
ethyl lactate	1.86E−01	Ta	0.00E+00
TMAH	2.01E−02	Cu ECP additive X	5.53E−04
TDEAH	0.00E+00	Cu ECP Additive Y	2.74E−04
TDMAS	0.00E+00	$Na_2B_4O_7$	8.69E−03
DEA	0.00E+00		
PDMAT	0.00E+00		

Table A.10 CMOS manufacturing inventory per wafer, 130 nm node

130 nm		130 nm	
Inputs	g/wafer	Inputs	g/wafer
GWG		*Flammable*	
CF_4	7.94E−03	SiH_4	1.21E−02
CHF_3	1.39E−03	H_2	9.73E−02
C_2F_6	9.80E−04	DCS	1.73E−03
CH_4	5.60E−01	B_2H_6	0.00E+00
CO_2	0.00E+00	C_4F_8	2.25E−03
NF_3	1.46E−01	GeH_4	3.92E−03
C_4F_6	4.34E−04	C_2H_2	0.00E+00
N_2O	0.00E+00	C_2H_4	0.00E+00
Hazardous			
F_2	0.00E+00	*Other chemicals*	
HF (gas)	0.00E+00	O_2	1.36E+00
SiF_4	2.84E−03	Ar	3.61E+00
C_4F_8	7.95E−04	N_2	1.47E+00
C_2F_4	0.00E+00	CMP polishing solution	1.67E+01
COF_2	0.00E+00	H_2O_2	1.84E+01
HCl (gas)	8.41E−02	abrasive	5.27E−01
Cl_2	1.71E−03	silica slurry	5.27E−01
$SiCl_4$	0.00E+00	benzotriazole	2.17E−02
PH_3	3.01E−04	CMP surfactants	4.05E−01
BTBAS	1.29E−03	citric acid	3.46E−02
AsH_3	0.00E+00	tungsten CMP slurry	2.78E+00
BCl_3	0.00E+00	Al	0.00E+00
Pb	1.35E−05	As	0.00E+00
WF_6	3.27E−02	Cu	1.10E−03
HBr	7.97E−03	Pt	0.00E+00
Br_2	0.00E+00	Ti	1.14E−07
HMDS	2.04E−01	He	8.99E−04
OMCTS	0.00E+00	NH_3	1.39E−02

Table A.11 CMOS manufacturing inventory per wafer, 130 nm node

130 nm		130 nm	
Inputs	g/wafer	Inputs	g/wafer
Criteria		*Other chemicals*	
CO	0.00E+00	$CuSO_4$	1.75E−02
NO	0.00E+00	ArH	0.00E+00
NO_2	0.00E+00	100% HF liquid	3.47E−01
SO_2	0.00E+00	H_2SO_4	1.92E+02
N_2O	6.96E−03	HCl	4.54E+00
O_3	0.00E+00	NH_4OH	5.99E+00
		IPA	1.58E+00
VOC		oxide CMP slurry, chemicals	5.33E+00
DMA	0.00E+00	BF_3	0.00E+00
MMA	0.00E+00	Cr	0.00E+00
TDMAT	1.21E−03	Au	1.64E−04
TMS	0.00E+00	Sn	3.74E−06
TEOS	0.00E+00	Co	8.76E−05
formaldehyde	2.42E−02	Ni	7.36E−03
m-cresol	1.54E−02	polyimide laminate	2.49E−03
p-cresol	1.52E−02	laminate solvent	0.00E+00
PGMEA	1.96E−01	H_3PO_4	3.61E−02
n-methyl-2-pyrollidone	1.12E−02	$CuCl_2$	1.00E−05
PGME	2.57E−01	W	7.74E−06
ethyl lactate	2.46E−01	Ta	1.95E+00
TMAH	2.65E−02	Cu ECP additive X	5.10E−03
TDEAH	0.00E+00	Cu ECP Additive Y	2.53E−03
TDMAS	0.00E+00	$Na_2B_4O_7$	1.15E−02
DEA	0.00E+00		
PDMAT	0.00E+00		

Table A.12 CMOS manufacturing inventory per wafer, 90 nm node

90 nm		90 nm	
Inputs	g/wafer	Inputs	g/wafer
GWG		*Flammable*	
CF_4	1.18E−02	SiH_4	1.76E−02
CHF_3	3.38E−03	H_2	9.75E−02
C_2F_6	0.00E+00	DCS	6.93E−03
CH_4	6.91E−01	B_2H_6	2.37E−03
CO_2	0.00E+00	C_4F_8	1.07E−03
NF_3	1.64E−01	GeH_4	3.92E−03
C_4F_6	1.45E−04	C_2H_2	0.00E+00
N_2O	0.00E+00	C_2H_4	0.00E+00
Hazardous			
F_2	0.00E+00	*Other chemicals*	
HF (gas)	0.00E+00	O_2	3.13E+00
SiF_4	9.40E−04	Ar	1.98E+00
C_4F_8	7.95E−04	N_2	1.80E+00

(continued)

Table A.12 (continued)

90 nm		90 nm	
Inputs	g/wafer	Inputs	g/wafer
C_2F_4	0.00E+00	CMP polishing solution	1.95E+01
COF_2	0.00E+00	H_2O_2	1.82E+01
HCl (gas)	8.41E−02	abrasive	6.15E−01
Cl_2	1.71E−03	silica slurry	6.15E−01
$SiCl_4$	0.00E+00	benzotriazole	2.53E−02
PH_3	3.01E−04	CMP surfactants	4.63E−01
BTBAS	1.29E−03	citric acid	3.50E−02
AsH_3	0.00E+00	tungsten CMP slurry	2.78E+00
BCl_3	0.00E+00	Al	0.00E+00
Pb	1.35E−05	As	0.00E+00
WF_6	3.27E−02	Cu	1.18E−03
HBr	7.97E−03	Pt	0.00E+00
Br_2	0.00E+00	Ti	1.60E−04
HMDS	2.59E−01	He	1.38E−01
OMCTS	4.69E−03	NH_3	1.62E−02

Table A.13 CMOS manufacturing inventory per wafer, 90 nm node

90 nm		90 nm	
Inputs	g/wafer	Inputs	g/wafer
Criteria		*Other chemicals*	
CO	0.00E+00	$CuSO_4$	2.00E−02
NO	0.00E+00	ArH	0.00E+00
NO_2	0.00E+00	100% HF liquid	3.33E−01
SO_2	0.00E+00	H_2SO_4	1.83E+02
N_2O	1.39E−02	HCl	4.73E+00
O_3	0.00E+00	NH_4OH	6.23E+00
		IPA	1.19E+00
VOC		oxide CMP slurry, chemicals	5.33E+00
DMA	0.00E+00	BF_3	0.00E+00
MMA	0.00E+00	Cr	0.00E+00
TDMAT	1.82E−03	Au	1.64E−04
TMS	2.94E−03	Sn	3.74E−06
TEOS	1.40E−03	Co	8.76E−05
formaldehyde	3.00E−02	Ni	7.36E−03
m-cresol	3.60E−02	polyimide laminate	2.49E−03
p-cresol	1.80E−02	laminate solvent	0.00E+00
PGMEA	2.55E−01	H_3PO_4	2.43E−02
n-methyl-2-pyrollidone	3.35E−02	$CuCl_2$	1.05E−03
PGME	3.06E−01	W	6.74E−03
ethyl lactate	2.90E−01	Ta	1.37E+00
TMAH	3.13E−02	Cu ECP additive X	6.13E−03
TDEAH	0.00E+00	Cu ECP Additive Y	2.89E−03
TDMAS	0.00E+00	$Na_2B_4O_7$	2.55E−01
DEA	0.00E+00		
PDMAT	4.25E−03		

Table A.14 CMOS manufacturing inventory per wafer, 65 nm node

65 nm		65 nm	
Inputs	g/wafer	Inputs	g/wafer
GWG		*Flammable*	
CF_4	1.37E−02	SiH_4	1.70E−02
CHF_3	3.30E−03	H_2	1.02E−01
C_2F_6	0.00E+00	DCS	9.81E−03
CH_4	9.84E−01	B_2H_6	2.98E−03
CO_2	0.00E+00	C_4F_8	1.46E−03
NF_3	1.57E−01	GeH_4	7.40E−03
C_4F_6	2.05E−04	C_2H_2	0.00E+00
N_2O	0.00E+00	C_2H_4	0.00E+00
Hazardous			
F_2	0.00E+00	*Other chemicals*	
HF (gas)	0.00E+00	O_2	3.91E+00
SiF_4	3.07E−04	Ar	1.50E+00
C_4F_8	7.50E−04	N_2	1.65E+00
C_2F_4	0.00E+00	CMP polishing solution	2.10E+01
COF_2	0.00E+00	H_2O_2	1.93E+01
HCl (gas)	1.59E−01	abrasive	6.63E−01
Cl_2	2.15E−03	silica slurry	6.63E−01
$SiCl_4$	0.00E+00	benzotriazole	2.73E−02
PH_3	2.84E−04	CMP surfactants	4.91E−01
BTBAS	1.21E−03	citric acid	3.34E−02
AsH_3	0.00E+00	tungsten CMP slurry	2.62E+00
BCl_3	0.00E+00	Al	0.00E+00
Pb	0.00E+00	As	0.00E+00
WF_6	3.09E−02	Cu	1.28E−03
HBr	1.00E−02	Pt	0.00E+00
Br_2	0.00E+00	Ti	3.03E−05
HMDS	2.71E−01	He	3.08E−01
OMCTS	5.90E−03	NH_3	1.77E−02

Table A.15 CMOS manufacturing inventory per wafer, 65 nm node

65 nm		65 nm	
Inputs	g/wafer	Inputs	g/wafer
Criteria		*Other chemicals*	
CO	0.00E+00	$CuSO_4$	2.12E−02
NO	0.00E+00	ArH	0.00E+00
NO_2	0.00E+00	100% HF liquid	2.13E−01
SO_2	0.00E+00	H_2SO_4	2.08E+02
N_2O	1.31E−02	HCl	4.46E+00
O_3	4.85E−02	NH_4OH	5.87E+00
		IPA	1.13E+00
VOC		oxide CMP slurry, chemicals	5.03E+00

(continued)

Table A.15 (continued)

65 nm		65 nm	
Inputs	g/wafer	Inputs	g/wafer
DMA	0.00E+00	BF_3	0.00E+00
MMA	0.00E+00	Cr	0.00E+00
TDMAT	1.15E−03	Au	1.77E−04
TMS	3.00E−03	Sn	3.44E−03
TEOS	2.62E−02	Co	9.45E−05
formaldehyde	3.14E−02	Ni	7.97E−03
m-cresol	3.59E−02	polyimide laminate	2.35E−03
p-cresol	1.89E−02	laminate solvent	0.00E+00
PGMEA	2.66E−01	H_3PO_4	3.25E−02
n-methyl-2-pyrollidone	3.16E−02	$CuCl_2$	9.95E−04
PGME	3.22E−01	W	8.48E−03
ethyl lactate	3.05E−01	Ta	1.83E+00
TMAH	3.30E−02	Cu ECP additive X	6.47E−03
TDEAH	0.00E+00	Cu ECP Additive Y	3.07E−03
TDMAS	0.00E+00	$Na_2B_4O_7$	2.42E−01
DEA	0.00E+00		
PDMAT	9.35E−03		

Table A.16 CMOS manufacturing inventory per wafer, 45 nm node

45 nm		45 nm	
Inputs	g/wafer	Inputs	g/wafer
GWG		*Flammable*	
CF_4	3.46E−02	SiH_4	1.68E−02
CHF_3	3.30E−03	H_2	1.02E−01
C_2F_6	0.00E+00	DCS	3.27E−03
CH_4	1.01E+00	B_2H_6	0.00E+00
CO_2	0.00E+00	C_4F_8	0.00E+00
NF_3	1.76E−01	GeH_4	7.40E−03
C_4F_6	2.05E−04	C_2H_2	0.00E+00
N_2O	0.00E+00	C_2H_4	0.00E+00
Hazardous			
F_2	0.00E+00	*Other chemicals*	
HF (gas)	0.00E+00	O_2	4.31E+00
SiF_4	4.09E−04	Ar	1.92E+00
C_4F_8	7.50E−04	N_2	1.66E+00
C_2F_4	0.00E+00	CMP polishing solution	2.36E+01
COF_2	0.00E+00	H_2O_2	2.33E+01
HCl (gas)	1.59E−01	abrasive	7.46E−01
Cl_2	2.35E−03	silica slurry	7.46E−01
$SiCl_4$	0.00E+00	benzotriazole	3.07E−02

(continued)

Table A.16 (continued)

45 nm		45 nm	
Inputs	g/wafer	Inputs	g/wafer
PH_3	2.84E−04	CMP surfactants	5.45E−01
BTBAS	6.07E−04	citric acid	3.38E−02
AsH_3	7.10E−05	tungsten CMP slurry	2.62E+00
BCl_3	0.00E+00	Al	0.00E+00
Pb	0.00E+00	As	0.00E+00
WF_6	3.09E−02	Cu	1.53E−03
HBr	1.00E−02	Pt	0.00E+00
Br_2	0.00E+00	Ti	1.07E−07
HMDS	2.88E−01	He	3.21E−01
OMCTS	5.90E−03	NH_3	1.22E−02

Table A.17 CMOS manufacturing inventory per wafer, 45 nm node

45 nm		45 nm	
Inputs	g/wafer	Inputs	g/wafer
Criteria		*Other chemicals*	
CO	7.58E−05	$CuSO_4$	2.36E−02
NO	0.00E+00	ArH	0.00E+00
NO_2	0.00E+00	100% HF liquid	1.15E−01
SO_2	0.00E+00	H_2SO_4	2.53E+02
N_2O	1.31E−02	HCl	5.35E+00
O_3	4.85E−02	NH_4OH	6.97E+00
		IPA	1.11E+00
VOC		oxide CMP slurry, chemicals	5.03E+00
DMA	0.00E+00	BF_3	0.00E+00
MMA	0.00E+00	Cr	0.00E+00
TDMAT	1.15E−03	Au	2.26E−04
TMS	3.00E−03	Sn	4.38E−03
TEOS	2.62E−02	Co	1.20E−04
formaldehyde	3.35E−02	Ni	1.01E−02
m-cresol	3.72E−02	polyimide laminate	2.35E−03
p-cresol	2.02E−02	laminate solvent	0.00E+00
PGMEA	2.82E−01	H_3PO_4	0.00E+00
n-methyl-2-pyrollidone	3.16E−02	$CuCl_2$	9.96E−04
PGME	3.44E−01	W	4.81E−07
ethyl lactate	3.27E−01	Ta	0.00E+00
TMAH	3.52E−02	Cu ECP additive X	7.16E−03
TDEAH	5.35E−05	Cu ECP Additive Y	3.41E−03
TDMAS	0.00E+00	$Na_2B_4O_7$	2.43E−01
DEA	0.00E+00		
PDMAT	1.20E−02		

Table A.18 CMOS manufacturing inventory per wafer, 32 nm node

32 nm		32 nm	
Inputs	g/wafer	Inputs	g/wafer
GWG		*Flammable*	
CF_4	2.63E−02	SiH_4	8.29E−03
CHF_3	9.17E−04	H_2	4.72E−02
C_2F_6	0.00E+00	DCS	1.44E−03
CH_4	9.35E−01	B_2H_6	0.00E+00
CO_2	0.00E+00	C_4F_8	0.00E+00
NF_3	1.74E−01	GeH_4	3.26E−03
C_4F_6	1.80E−04	C_2H_2	0.00E+00
N_2O	0.00E+00	C_2H_4	0.00E+00
Hazardous			
F_2	0.00E+00	*Other chemicals*	
HF (gas)	0.00E+00	O_2	4.28E+00
SiF_4	5.40E−04	Ar	2.32E+00
C_4F_8	3.30E−04	N_2	1.56E+00
C_2F_4	0.00E+00	CMP polishing solution	2.54E+01
COF_2	0.00E+00	H_2O_2	2.29E+01
HCl (gas)	6.98E−02	abrasive	8.02E−01
Cl_2	2.82E−03	silica slurry	8.02E−01
$SiCl_4$	0.00E+00	benzotriazole	3.30E−02
PH_3	2.50E−04	CMP surfactants	5.76E−01
BTBAS	5.34E−04	citric acid	3.04E−02
AsH_3	0.00E+00	tungsten CMP slurry	2.31E+00
BCl_3	1.06E−03	Al	0.00E+00
Pb	0.00E+00	As	0.00E+00
WF_6	1.36E−02	Cu	1.25E−03
HBr	8.82E−03	Pt	0.00E+00
Br_2	0.00E+00	Ti	2.65E−05
HMDS	3.69E−01	He	2.95E−01
OMCTS	5.19E−03	NH_3	1.06E−02

Table A.19 CMOS manufacturing inventory per wafer, 32 nm node

32 nm		32 nm	
Inputs	g/wafer	Inputs	g/wafer
Criteria		*Other chemicals*	
CO	6.67E−05	$CuSO_4$	2.28E−02
NO	0.00E+00	ArH	0.00E+00
NO_2	0.00E+00	100% HF liquid	1.08E−01
SO_2	0.00E+00	H_2SO_4	3.25E+02
N_2O	5.78E−03	HCl	2.20E+00
O_3	4.27E−02	NH_4OH	3.24E+00
		IPA	6.50E−01
VOC		oxide CMP slurry, chemicals	4.43E+00

(continued)

Table A.19   (continued)

32 nm		32 nm	
Inputs	g/wafer	Inputs	g/wafer
DMA	0.00E+00	BF$_3$	0.00E+00
MMA	0.00E+00	Cr	0.00E+00
TDMAT	1.01E−03	Au	3.14E−04
TMS	8.12E−04	Sn	6.08E−03
TEOS	2.19E−02	Co	1.68E−04
formaldehyde	4.39E−02	Ni	1.41E−02
m-cresol	2.78E−02	polyimide laminate	2.07E−03
p-cresol	2.75E−02	laminate solvent	0.00E+00
PGMEA	3.55E−01	H$_3$PO$_4$	0.00E+00
n-methyl-2-pyrollidone	9.27E−03	CuCl$_2$	1.31E−05
PGME	4.66E−01	W	0.00E+00
ethyl lactate	4.45E−01	Ta	0.00E+00
TMAH	4.80E−02	Cu ECP additive X	6.66E−03
TDEAH	4.71E−05	Cu ECP Additive Y	3.30E−03
TDMAS	0.00E+00	Na$_2$B$_4$O$_7$	2.08E−02
DEA	0.00E+00		
PDMAT	1.06E−02		

A.2   CMOS Logic Manufacturing Emissions Data

Table A.20   CMOS Manufacturing emissions per wafer, 350 nm node

350 nm	Emissions to air		
kg/wafer	Expected value	Low	High
GWG			
CF$_4$	1.09E−03	2.37E−02	9.48E−02
CHF$_3$	7.18E−05	4.35E−04	1.74E−03
C$_2$F$_6$	0.00E+00	0.00E+00	0.00E+00
CH$_4$	4.90E−05	2.18E−01	8.73E−01
CO$_2$	1.20E+00	0.00E+00	0.00E+00
NF$_3$	2.88E−05	1.01E−02	4.05E−02
C$_4$F$_6$	0.00E+00	0.00E+00	0.00E+00
N$_2$O	0.00E+00	0.00E+00	0.00E+00
Hazardous			
F$_2$	2.76E−03	0.00E+00	0.00E+00
HF (gas)	1.53E−03	0.00E+00	0.00E+00
SiF$_4$	1.04E−05	2.21E−05	8.82E−05
C$_4$F$_8$	2.94E−07	1.62E−04	6.47E−04
C$_2$F$_4$	3.47E−06	0.00E+00	0.00E+00
COF$_2$	7.79E−06	0.00E+00	0.00E+00
HCl (gas)	6.40E−07	0.00E+00	0.00E+00
Cl$_2$	2.15E−04	2.69E−03	1.08E−02

(continued)

Table A.20 (continued)

350 nm	Emissions to air		
kg/wafer	Expected value	Low	High
$SiCl_4$	4.40E−06	0.00E+00	0.00E+00
BTBAS	8.25E−06	5.23E−04	2.09E−03
AsH_3	0.00E+00	0.00E+00	0.00E+00
BCl_3	8.96E−05	3.13E−03	1.25E−02
Pb	9.90E−06	5.50E−06	2.20E−05
WF_6	1.06E−02	3.99E−02	1.60E−01
HBr	0.00E+00	2.16E−03	8.65E−03
Br_2	0.00E+00	0.00E+00	0.00E+00
HMDS	0.00E+00	6.78E−02	2.71E−01
OMCTS	0.00E+00	0.00E+00	0.00E+00
Criteria			
CO	2.53E−03	0.00E+00	0.00E+00
NO	2.94E−05	0.00E+00	0.00E+00
NO_2	0.00E+00	0.00E+00	0.00E+00
SO_2	0.00E+00	0.00E+00	0.00E+00
N_2O	1.93E−03	1.79E−02	7.18E−02
O_3	0.00E+00	0.00E+00	0.00E+00

Table A.21 CMOS Manufacturing emissions per wafer, 350 nm node

350 nm	Emissions to air		
kg/wafer	Expected value	Low	High
VOC			
DMA	1.04E−06	0.00E+00	0.00E+00
MMA	2.40E−05	0.00E+00	0.00E+00
TDMAT	0.00E+00	2.47E−04	9.88E−04
TMS	0.00E+00	0.00E+00	0.00E+00
TEOS	0.00E+00	0.00E+00	0.00E+00
formaldehyde (CH_2O)	7.98E−05	8.06E−03	3.22E−02
m-cresol	5.06E−06	5.11E−03	2.04E−02
p-cresol	5.01E−06	5.06E−03	2.03E−02
PGMEA	9.39E−03	6.52E−02	2.61E−01
n-Methyl-2-Pyrollidone	0.00E+00	4.54E−03	1.82E−02
PGME	6.97E−03	8.56E−02	3.42E−01
ethyl lactate	8.02E−03	8.18E−02	3.27E−01
TMAH	8.82E−05	8.82E−03	3.53E−02
TDEAH	0.00E+00	0.00E+00	0.00E+00
TDMAS	0.00E+00	0.00E+00	0.00E+00
DEA	0.00E+00	0.00E+00	0.00E+00
PDMAT	0.00E+00	0.00E+00	0.00E+00
Flammable			
SiH_4	2.97E−04	4.19E−03	1.68E−02
H_2	9.19E−03	1.10E−01	4.39E−01
DCS	0.00E+00	0.00E+00	0.00E+00

(continued)

Table A.21 (continued)

350 nm	Emissions to air		
kg/wafer	Expected value	Low	High
B_2H_6	0.00E+00	0.00E+00	0.00E+00
C_4F_8	0.00E+00	0.00E+00	0.00E+00
GeH_4	0.00E+00	0.00E+00	0.00E+00
C_2H_2			
Other chemicals			
Ar	6.51E−01	3.28E−01	1.31E+00
N_2	4.49E−01	2.43E−01	9.72E−01
He	3.66E−04	1.83E−04	7.32E−04
O_2	3.88E−02	5.98E−02	2.39E−01
IPA	1.84E−02	2.87E+00	1.15E+01
NH_3	3.66E−04	1.83E−04	7.32E−04
Ti	8.33E−08	1.31E−05	5.22E−05
Cu	1.15E−07	7.77E−05	3.11E−04
Sn	2.74E−06	1.52E−06	6.08E−06

Table A.22 CMOS Manufacturing emissions per wafer, 250 nm node

250 nm	Emissions to air		
kg/wafer	Expected value	Low	High
GWG			
CF_4	1.28E−03	2.76E−02	1.10E−01
CHF_3	8.36E−05	4.71E−04	1.89E−03
C_2F_6	0.00E+00	0.00E+00	0.00E+00
CH_4	4.90E−05	2.38E−01	9.50E−01
CO_2	1.31E+00	0.00E+00	0.00E+00
NF_3	3.04E−05	1.05E−02	4.19E−02
C_4F_6	0.00E+00	0.00E+00	0.00E+00
N_2O	0.00E+00	0.00E+00	0.00E+00
Hazardous			
F_2	3.16E−03	0.00E+00	0.00E+00
HF (gas)	1.80E−03	0.00E+00	0.00E+00
SiF_4	1.07E−05	2.21E−05	8.82E−05
C_4F_8	2.94E−07	1.62E−04	6.47E−04
C_2F_4	3.47E−06	0.00E+00	0.00E+00
COF_2	8.21E−06	0.00E+00	0.00E+00
HCl (gas)	7.47E−07	0.00E+00	0.00E+00
Cl_2	2.49E−04	3.06E−03	1.22E−02
$SiCl_4$	5.13E−06	0.00E+00	0.00E+00
PH_3	0.00E+00	1.23E−04	4.90E−04
BTBAS	8.25E−06	5.23E−04	2.09E−03
AsH_3	0.00E+00	0.00E+00	0.00E+00
BCl_3	1.05E−04	3.65E−03	1.46E−02
Pb	9.90E−06	5.50E−06	2.20E−05
WF_6	1.23E−02	4.66E−02	1.86E−01

(continued)

Table A.22 (continued)

250 nm	Emissions to air		
kg/wafer	Expected value	Low	High
HBr	0.00E+00	2.16E−03	8.65E−03
Br$_2$	0.00E+00	0.00E+00	0.00E+00
HMDS	0.00E+00	7.53E−02	3.01E−01
OMCTS	0.00E+00	0.00E+00	0.00E+00
Criteria			
CO	2.95E−03	0.00E+00	0.00E+00
NO	2.94E−05	0.00E+00	0.00E+00
NO$_2$	0.00E+00	0.00E+00	0.00E+00
SO$_2$	0.00E+00	0.00E+00	0.00E+00
N$_2$O	2.21E−03	2.05E−02	8.19E−02
O$_3$	0.00E+00	0.00E+00	0.00E+00

Table A.23 CMOS Manufacturing emissions per wafer, 250 nm node

250 nm	Emissions to air		
kg/wafer	Expected value	Low	High
VOC			
DMA	2.09E−06	0.00E+00	0.00E+00
MMA	4.80E−05	0.00E+00	0.00E+00
TDMAT	0.00E+00	4.94E−04	1.98E−03
TMS	0.00E+00	0.00E+00	0.00E+00
TEOS	0.00E+00	0.00E+00	0.00E+00
formaldehyde	8.87E−05	8.96E−03	3.58E−02
m-cresol	5.62E−06	5.68E−03	2.27E−02
p-cresol	5.57E−06	5.63E−03	2.25E−02
PGMEA	1.04E−02	7.25E−02	2.90E−01
n-methyl-2-pyrollidone	0.00E+00	4.54E−03	1.82E−02
PGME	7.69E−03	9.51E−02	3.80E−01
ethyl lactate	8.91E−03	9.09E−02	3.64E−01
TMAH	9.80E−05	9.80E−03	3.92E−02
TDEAH	0.00E+00	0.00E+00	0.00E+00
TDMAS	0.00E+00	0.00E+00	0.00E+00
DEA	0.00E+00	0.00E+00	0.00E+00
PDMAT	0.00E+00	0.00E+00	0.00E+00
Flammable			
SiH$_4$	3.46E−04	4.37E−03	1.75E−02
H$_2$	1.07E−02	1.28E−01	5.12E−01
DCS	0.00E+00	0.00E+00	0.00E+00
B$_2$H$_6$	0.00E+00	0.00E+00	0.00E+00
C$_4$F$_8$	0.00E+00	0.00E+00	0.00E+00
GeH$_4$	0.00E+00	0.00E+00	0.00E+00
C$_2$H$_2$			

(continued)

Table A.23 (continued)

250 nm	Emissions to air		
kg/wafer	Expected value	Low	High
Other chemicals			
Ar	7.32E−01	3.68E−01	1.47E+00
N$_2$	6.05E−01	3.29E−01	1.32E+00
He	7.32E−04	3.66E−04	1.46E−03
O$_2$	4.02E−02	6.05E−02	2.42E−01
IPA	2.04E−02	3.03E+00	1.21E+01
NH$_3$	7.32E−04	3.66E−04	1.46E−03
Ti	8.33E−08	1.31E−05	5.22E−05
Cu	1.15E−07	1.33E−04	5.34E−04
Sn	2.74E−06	1.52E−06	6.08E−06

Table A.24 CMOS Manufacturing emissions per wafer, 180 nm node

180 nm	Emissions to air		
kg/wafer	Expected value	Low	High
GWG			
CF$_4$	2.80E−04	1.46E−03	5.83E−03
CHF$_3$	8.88E−05	4.73E−04	1.89E−03
C$_2$F$_6$	1.37E−05	2.97E−04	1.19E−03
CH$_4$	9.92E−05	1.95E−01	7.79E−01
CO$_2$	1.01E+00	0.00E+00	0.00E+00
NF$_3$	2.07E−04	4.98E−02	1.99E−01
C$_4$F$_6$	0.00E+00	0.00E+00	0.00E+00
N$_2$O	0.00E+00	0.00E+00	0.00E+00
Hazardous			
F$_2$	3.67E−03	0.00E+00	0.00E+00
HF (gas)	1.67E−03	0.00E+00	0.00E+00
SiF$_4$	2.51E−05	1.01E−03	4.05E−03
C$_4$F$_8$	2.74E−07	1.51E−04	6.03E−04
C$_2$F$_4$	8.71E−06	0.00E+00	0.00E+00
COF$_2$	1.18E−05	0.00E+00	0.00E+00
HCl (gas)	7.95E−07	0.00E+00	0.00E+00
Cl$_2$	2.64E−04	3.20E−03	1.28E−02
SiCl$_4$	5.46E−06	0.00E+00	0.00E+00
PH$_3$	0.00E+00	1.14E−04	4.57E−04
BTBAS	7.69E−06	4.87E−04	1.95E−03
AsH$_3$	0.00E+00	0.00E+00	0.00E+00
BCl$_3$	1.11E−04	3.88E−03	1.55E−02
Pb	9.22E−06	5.12E−06	2.05E−05
WF$_6$	1.31E−02	4.96E−02	1.98E−01

(continued)

Table A.24 (continued)

180 nm	Emissions to air		
kg/wafer	Expected value	Low	High
HBr	0.00E+00	2.01E−03	8.06E−03
Br$_2$	0.00E+00	0.00E+00	0.00E+00
HMDS	0.00E+00	7.72E−02	3.09E−01
OMCTS	0.00E+00	0.00E+00	0.00E+00
Criteria			
CO	4.86E−02	0.00E+00	0.00E+00
NO	2.09E−03	0.00E+00	0.00E+00
NO$_2$	7.66E−04	0.00E+00	0.00E+00
SO$_2$	0.00E+00	0.00E+00	0.00E+00
N$_2$O	2.01E−04	2.64E−03	1.05E−02
O$_3$	0.00E+00	0.00E+00	0.00E+00

Table A.25 CMOS Manufacturing emissions per wafer, 180 nm node

180 nm	Emissions to air		
kg/wafer	Expected value	Low	High
VOC			
DMA	1.94E−06	0.00E+00	0.00E+00
MMA	4.47E−05	0.00E+00	0.00E+00
TDMAT	0.00E+00	4.60E−04	1.84E−03
TMS	0.00E+00	0.00E+00	0.00E+00
TEOS	0.00E+00	0.00E+00	0.00E+00
formaldehyde	9.09E−05	9.18E−03	3.67E−02
m-cresol	5.76E−06	5.82E−03	2.33E−02
p-cresol	5.71E−06	5.76E−03	2.31E−02
PGMEA	1.07E−02	7.43E−02	2.97E−01
n-methyl-2-pyrollidone	0.00E+00	4.23E−03	1.69E−02
PGME	7.84E−03	9.74E−02	3.90E−01
ethyl lactate	9.13E−03	9.31E−02	3.72E−01
TMAH	1.00E−04	1.00E−02	4.02E−02
TDEAH	0.00E+00	0.00E+00	0.00E+00
TDMAS	0.00E+00	0.00E+00	0.00E+00
DEA	0.00E+00	0.00E+00	0.00E+00
PDMAT	0.00E+00	0.00E+00	0.00E+00
Flammable			
SiH$_4$	5.43E−04	4.65E−03	1.86E−02
H$_2$	1.14E−02	1.36E−01	5.45E−01
DCS	0.00E+00	0.00E+00	0.00E+00
B$_2$H$_6$	0.00E+00	0.00E+00	0.00E+00
C$_4$F$_8$	2.74E−05	0.00E+00	0.00E+00
GeH$_4$	0.00E+00	0.00E+00	0.00E+00
C$_2$H$_2$			

(continued)

Table A.25 (continued)

180 nm	Emissions to air		
kg/wafer	Expected value	Low	High
Other chemicals			
Ar	2.66E+00	1.33E+00	5.33E+00
N_2	5.93E−01	3.21E−01	1.29E+00
He	6.82E−04	3.41E−04	1.36E−03
O_2	3.40E−02	4.12E−01	1.65E+00
IPA	2.09E−02	2.67E+00	1.07E+01
NH_3	6.82E−04	3.41E−04	1.36E−03
Ti	7.76E−08	1.22E−05	4.87E−05
Cu	1.07E−07	1.65E−04	6.61E−04
Sn	2.55E−06	1.42E−06	5.66E−06

Table A.26 CMOS Manufacturing emissions per wafer, 130 nm node

130 nm	Emissions to air		
kg/wafer	Expected value	Low	High
GWG			
CF_4	7.32E−04	3.97E−03	1.59E−02
CHF_3	6.34E−06	6.97E−04	2.79E−03
C_2F_6	2.64E−05	4.90E−04	1.96E−03
CH_4	5.95E−04	2.80E−01	1.12E+00
CO_2	1.37E+00	0.00E+00	0.00E+00
NF_3	2.89E−04	7.29E−02	2.92E−01
C_4F_6	3.40E−05	2.17E−04	8.67E−04
N_2O	0.00E+00	0.00E+00	0.00E+00
Hazardous			
F_2	1.75E−03	0.00E+00	0.00E+00
HF (gas)	5.46E−03	0.00E+00	0.00E+00
SiF_4	5.29E−05	1.42E−03	5.67E−03
C_4F_8	3.61E−06	3.98E−04	1.59E−03
C_2F_4	1.76E−05	0.00E+00	0.00E+00
COF_2	2.08E−05	0.00E+00	0.00E+00
HCl (gas)	0.00E+00	4.20E−02	1.68E−01
Cl_2	1.87E−05	8.57E−04	3.43E−03
$SiCl_4$	0.00E+00	0.00E+00	0.00E+00
PH_3	0.00E+00	1.51E−04	6.02E−04
BTBAS	1.01E−05	6.43E−04	2.57E−03
AsH_3	0.00E+00	0.00E+00	0.00E+00
BCl_3	0.00E+00	0.00E+00	0.00E+00
Pb	1.22E−05	6.76E−06	2.70E−05
WF_6	4.32E−03	1.64E−02	6.54E−02

(continued)

Table A.26   (continued)

130 nm	Emissions to air		
kg/wafer	Expected value	Low	High
HBr	0.00E+00	3.99E−03	1.59E−02
Br$_2$	0.00E+00	0.00E+00	0.00E+00
HMDS	0.00E+00	1.02E−01	4.07E−01
OMCTS	0.00E+00	0.00E+00	0.00E+00
Criteria			
CO	4.22E−03	0.00E+00	0.00E+00
NO	2.21E−03	0.00E+00	0.00E+00
NO$_2$	2.88E−04	0.00E+00	0.00E+00
SO$_2$	0.00E+00	0.00E+00	0.00E+00
N$_2$O	2.65E−04	3.48E−03	1.39E−02
O$_3$	0.00E+00	0.00E+00	0.00E+00

Table A.27   CMOS Manufacturing emissions per wafer, 130 nm node

130 nm	Emissions to Air		
kg/wafer	Expected value	Low	High
VOC			
DMA	2.57E−06	0.00E+00	0.00E+00
MMA	5.90E−05	0.00E+00	0.00E+00
TDMAT	0.00E+00	6.07E−04	2.43E−03
TMS	0.00E+00	0.00E+00	0.00E+00
TEOS	0.00E+00	0.00E+00	0.00E+00
formaldehyde	1.20E−04	1.21E−02	4.84E−02
m-cresol	7.60E−06	7.68E−03	3.07E−02
p-cresol	7.53E−06	7.60E−03	3.04E−02
PGMEA	1.41E−02	9.80E−02	3.92E−01
n-methyl-2-pyrollidone	0.00E+00	5.58E−03	2.23E−02
PGME	1.03E−02	1.29E−01	5.14E−01
ethyl lactate	1.20E−02	1.23E−01	4.91E−01
TMAH	1.33E−04	1.33E−02	5.30E−02
TDEAH	0.00E+00	0.00E+00	0.00E+00
TDMAS	0.00E+00	0.00E+00	0.00E+00
DEA	0.00E+00	0.00E+00	0.00E+00
PDMAT	0.00E+00	0.00E+00	0.00E+00
Flammable			
SiH$_4$	3.56E−04	6.04E−03	2.42E−02
H$_2$	3.76E−03	4.86E−02	1.95E−01
DCS	0.00E+00	8.67E−04	3.47E−03
B$_2$H$_6$	0.00E+00	0.00E+00	0.00E+00
C$_4$F$_8$	2.25E−03	1.13E−03	4.50E−03
GeH$_4$	0.00E+00	1.96E−03	7.84E−03
C$_2$H$_2$			

(continued)

Table A.27   (continued)

130 nm	Emissions to Air		
kg/wafer	Expected value	Low	High
Other chemicals			
Ar	3.60E+00	1.80E+00	7.22E+00
N_2	1.41E+00	7.37E−01	2.95E+00
He	8.99E−04	4.50E−04	1.80E−03
O_2	4.79E−02	6.82E−01	2.73E+00
IPA	2.75E−02	7.92E−01	3.17E+00
NH_3	8.99E−04	4.50E−04	1.80E−03
Ti	1.02E−07	5.69E−08	2.28E−07
Cu	1.41E−07	5.48E−04	2.19E−03
Sn	3.36E−06	1.87E−06	7.47E−06

Table A.28   CMOS Manufacturing emissions per wafer, 90 nm node

90 nm	Emissions to Air		
kg/wafer	Expected value	Low	High
GWG			
CF_4	4.58E−04	5.91E−03	2.36E−02
CHF_3	1.21E−05	1.69E−03	6.77E−03
C_2F_6	3.86E−06	0.00E+00	0.00E+00
CH_4	1.29E−03	3.45E−01	1.38E+00
CO_2	1.68E+00	0.00E+00	0.00E+00
NF_3	3.01E−04	8.18E−02	3.27E−01
C_4F_6	1.13E−05	7.23E−05	2.89E−04
N_2O	0.00E+00	0.00E+00	0.00E+00
Hazardous			
F_2	4.03E−03	0.00E+00	0.00E+00
HF (gas)	1.12E−02	0.00E+00	0.00E+00
SiF_4	5.46E−05	4.70E−04	1.88E−03
C_4F_8	1.69E−06	3.98E−04	1.59E−03
C_2F_4	8.53E−06	0.00E+00	0.00E+00
COF_2	5.45E−05	0.00E+00	0.00E+00
HCl (gas)	0.00E+00	4.20E−02	1.68E−01
Cl_2	1.87E−05	8.57E−04	3.43E−03
$SiCl_4$	0.00E+00	0.00E+00	0.00E+00
PH_3	0.00E+00	1.51E−04	6.02E−04
BTBAS	1.01E−05	6.43E−04	2.57E−03
AsH_3	0.00E+00	0.00E+00	0.00E+00
BCl_3	0.00E+00	0.00E+00	0.00E+00
Pb	1.22E−05	6.76E−06	2.70E−05
WF_6	4.32E−03	1.64E−02	6.54E−02

(continued)

Table A.28 (continued)

90 nm	Emissions to Air		
kg/wafer	Expected value	Low	High
HBr	0.00E+00	3.99E−03	1.59E−02
Br$_2$	0.00E+00	0.00E+00	0.00E+00
HMDS	0.00E+00	1.30E−01	5.18E−01
OMCTS	0.00E+00	2.34E−03	9.38E−03
Criteria			
CO	2.66E−02	0.00E+00	0.00E+00
NO	2.19E−03	0.00E+00	0.00E+00
NO$_2$	3.80E−04	0.00E+00	0.00E+00
SO$_2$	0.00E+00	0.00E+00	0.00E+00
N$_2$O	5.30E−04	6.96E−03	2.78E−02
O$_3$	0.00E+00	0.00E+00	0.00E+00

Table A.29 CMOS Manufacturing emissions per wafer, 90 nm node

90 nm	Emissions to Air		
kg/wafer	Expected value	Low	High
VOC			
DMA	3.85E−06	0.00E+00	0.00E+00
MMA	8.85E−05	0.00E+00	0.00E+00
TDMAT	0.00E+00	9.11E−04	3.64E−03
TMS	1.96E−03	1.47E−03	5.87E−03
TEOS	1.37E−03	6.98E−04	2.79E−03
formaldehyde	1.42E−04	1.50E−02	6.00E−02
m-cresol	8.98E−06	1.80E−02	7.19E−02
p-cresol	8.90E−06	8.99E−03	3.59E−02
PGMEA	1.67E−02	1.27E−01	5.10E−01
n-methyl-2-pyrollidone	0.00E+00	1.68E−02	6.70E−02
PGME	1.21E−02	1.53E−01	6.13E−01
ethyl lactate	1.42E−02	1.45E−01	5.81E−01
TMAH	1.57E−04	1.57E−02	6.27E−02
TDEAH	0.00E+00	0.00E+00	0.00E+00
TDMAS	0.00E+00	0.00E+00	0.00E+00
DEA	0.00E+00	0.00E+00	0.00E+00
PDMAT	0.00E+00	2.12E−03	8.50E−03
Flammable			
SiH$_4$	4.11E−04	8.78E−03	3.51E−02
H$_2$	3.76E−03	4.88E−02	1.95E−01
DCS	5.20E−03	3.47E−03	1.39E−02
B$_2$H$_6$	1.72E−03	1.18E−03	4.73E−03
C$_4$F$_8$	1.08E−03	5.33E−04	2.13E−03
GeH$_4$	0.00E+00	1.96E−03	7.84E−03
C$_2$H$_2$			

(continued)

Table A.29 (continued)

90 nm	Emissions to Air		
kg/wafer	Expected value	Low	High
Other chemicals			
Ar	1.97E+00	9.92E−01	3.97E+00
N$_2$	1.68E+00	9.00E−01	3.60E+00
He	1.35E−03	6.92E−02	2.77E−01
O$_2$	3.62E−02	1.56E+00	6.25E+00
IPA	3.25E−02	5.95E−01	2.38E+00
NH$_3$	1.35E−03	6.92E−02	2.77E−01
Ti	1.02E−07	8.00E−05	3.20E−04
Cu	1.41E−07	5.92E−04	2.37E−03
Sn	3.36E−06	1.87E−06	7.47E−06

Table A.30 CMOS Manufacturing emissions per wafer, 65 nm node

65 nm	Emissions to Air		
kg/wafer	Expected value	Low	High
GWG			
CF$_4$	1.89E−03	9.44E−04	3.78E−03
CHF$_3$	1.15E−05	5.73E−06	2.29E−05
C$_2$F$_6$	1.18E−05	5.91E−06	2.36E−05
CH$_4$	4.92E−03	2.46E−03	9.83E−03
CO$_2$	1.61E+00	8.06E−01	3.22E+00
NF$_3$	4.06E−04	2.03E−04	8.13E−04
C$_4$F$_6$	4.27E−05	2.14E−05	8.55E−05
N$_2$O	2.50E−04	1.25E−04	5.00E−04
Hazardous			
F$_2$	1.08E−03	5.38E−04	2.15E−03
HF (gas)	3.48E−02	1.74E−02	6.96E−02
SiF$_4$	9.73E−05	4.86E−05	1.95E−04
C$_4$F$_8$	4.32E−06	2.16E−06	8.64E−06
C$_2$F$_4$	8.05E−06	4.02E−06	1.61E−05
COF$_2$	1.33E−05	6.64E−06	2.66E−05
HCl (gas)	0.00E+00	0.00E+00	0.00E+00
Cl$_2$	2.35E−05	1.17E−05	4.69E−05
SiCl$_4$	0.00E+00	0.00E+00	0.00E+00
PH$_3$	0.00E+00	0.00E+00	0.00E+00
BTBAS	4.78E−06	2.39E−06	9.57E−06
AsH$_3$	0.00E+00	0.00E+00	0.00E+00
BCl$_3$	0.00E+00	0.00E+00	0.00E+00
Pb	0.00E+00	0.00E+00	0.00E+00
WF$_6$	2.04E−03	1.02E−03	4.08E−03

(continued)

Table A.30 (continued)

65 nm	Emissions to Air		
kg/wafer	Expected value	Low	High
HBr	0.00E+00	0.00E+00	0.00E+00
Br$_2$	0.00E+00	0.00E+00	0.00E+00
HMDS	0.00E+00	0.00E+00	0.00E+00
OMCTS	0.00E+00	0.00E+00	0.00E+00
Criteria			
CO	6.00E−01	3.00E−01	1.20E+00
NO	6.78E−03	3.39E−03	1.36E−02
NO$_2$	6.46E−03	3.23E−03	1.29E−02
SO$_2$	0.00E+00	0.00E+00	0.00E+00
N$_2$O	2.50E−04	1.25E−04	5.00E−04
O$_3$	0.00E+00	0.00E+00	0.00E+00

Table A.31 CMOS Manufacturing emissions per wafer, 65 nm node

65 nm	Emissions to Air		
kg/wafer	Expected value	Low	High
VOC			
DMA	1.21E−06	6.05E−07	2.42E−06
MMA	2.78E−05	1.39E−05	5.57E−05
TDMAT	0.00E+00	0.00E+00	0.00E+00
TMS	2.05E−06	1.02E−06	4.09E−06
TEOS	1.88E−07	9.38E−08	3.75E−07
formaldehyde	1.49E−04	7.45E−05	2.98E−04
m-cresol	9.45E−06	4.73E−06	1.89E−05
p-cresol	9.36E−06	4.68E−06	1.87E−05
PGMEA	1.75E−02	8.77E−03	3.51E−02
n-methyl-2-pyrollidone	0.00E+00	0.00E+00	0.00E+00
PGME	1.27E−02	6.35E−03	2.54E−02
ethyl lactate	1.50E−02	7.49E−03	2.99E−02
TMAH	1.65E−04	8.24E−05	3.30E−04
TDEAH	0.00E+00	0.00E+00	0.00E+00
TDMAS	0.00E+00	0.00E+00	0.00E+00
DEA	0.00E+00	0.00E+00	0.00E+00
PDMAT	0.00E+00	0.00E+00	0.00E+00
Flammable			
SiH$_4$	4.17E−04	2.09E−04	8.34E−04
H$_2$	3.24E−03	1.62E−03	6.49E−03
DCS	0.00E+00	0.00E+00	0.00E+00
B$_2$H$_6$	0.00E+00	0.00E+00	0.00E+00
C$_4$F$_8$	3.18E−04	1.59E−04	6.36E−04
GeH$_4$	0.00E+00	0.00E+00	0.00E+00
C$_2$H$_2$			

(continued)

Table A.31 (continued)

65 nm	Emissions to Air		
kg/wafer	Expected value	Low	High
Other chemicals			
Ar	1.77E+00	8.85E−01	3.54E+00
N_2	1.25E+00	6.25E−01	2.50E+00
He	1.35E−01	6.75E−02	2.70E−01
O_2	4.51E−02	2.26E−02	9.03E−02
IPA	3.42E−02	1.71E−02	6.84E−02
NH_3	1.35E−01	6.75E−02	2.70E−01
Ti	0.00E+00	0.00E+00	0.00E+00
Cu	0.00E+00	0.00E+00	0.00E+00
Sn	0.00E+00	0.00E+00	0.00E+00

Table A.32 CMOS Manufacturing emissions per wafer, 45 nm node

45 nm	Emissions to air		
kg/wafer	Expected value	Low	High
GWG			
CF_4	2.14E−03	1.07E−03	4.28E−03
CHF_3	1.26E−05	6.30E−06	2.52E−05
C_2F_6	5.00E−06	2.50E−06	1.00E−05
CH_4	4.47E−03	2.23E−03	8.94E−03
CO_2	1.78E+00	8.89E−01	3.56E+00
NF_3	4.30E−04	2.15E−04	8.59E−04
C_4F_6	1.60E−05	8.01E−06	3.20E−05
N_2O	0.00E+00	2.50E−04	1.00E−03
Hazardous			
F_2	3.80E−03	1.90E−03	7.61E−03
HF (gas)	9.25E−03	4.62E−03	1.85E−02
SiF_4	8.87E−05	4.43E−05	1.77E−04
C_4F_8	2.05E−06	1.02E−06	4.09E−06
C_2F_4	8.05E−06	4.02E−06	1.61E−05
COF_2	5.66E−05	2.83E−05	1.13E−04
HCl (gas)	0.00E+00	0.00E+00	0.00E+00
Cl_2	2.35E−05	1.17E−05	4.69E−05
$SiCl_4$	0.00E+00	0.00E+00	0.00E+00
PH_3	2.84E−08	1.42E−08	5.68E−08
BTBAS	4.78E−06	2.39E−06	9.57E−06
AsH_3	3.55E−09	1.78E−09	7.10E−09
BCl_3	0.00E+00	0.00E+00	0.00E+00
Pb	0.00E+00	0.00E+00	0.00E+00
WF_6	4.08E−03	2.04E−03	8.16E−03

(continued)

Table A.32 (continued)

45 nm	Emissions to air		
kg/wafer	Expected value	Low	High
HBr	0.00E+00	0.00E+00	0.00E+00
Br$_2$	0.00E+00	0.00E+00	0.00E+00
HMDS	0.00E+00	0.00E+00	0.00E+00
OMCTS	0.00E+00	0.00E+00	0.00E+00
Criteria			
CO	6.29E−01	3.14E−01	1.26E+00
NO	6.98E−03	3.49E−03	1.40E−02
NO$_2$	6.69E−03	3.34E−03	1.34E−02
SO$_2$	0.00E+00	0.00E+00	0.00E+00
N$_2$O	5.00E−04	2.50E−04	1.00E−03
O$_3$	0.00E+00	0.00E+00	0.00E+00

Table A.33 CMOS Manufacturing emissions per wafer, 45 nm node

45 nm	Emissions to air		
kg/wafer	Expected value	Low	High
VOC			
DMA	2.42E−06	1.21E−06	4.84E−06
MMA	5.57E−05	2.78E−05	1.11E−04
TDMAT	0.00E+00	0.00E+00	0.00E+00
TMS	1.85E−03	9.26E−04	3.70E−03
TEOS	1.29E−03	6.45E−04	2.58E−03
formaldehyde	1.59E−04	7.97E−05	3.19E−04
m-cresol	1.01E−05	5.05E−06	2.02E−05
p-cresol	1.00E−05	5.00E−06	2.00E−05
PGMEA	1.87E−02	9.37E−03	3.75E−02
n-methyl-2-pyrollidone	0.00E+00	0.00E+00	0.00E+00
PGME	1.35E−02	6.77E−03	2.71E−02
ethyl lactate	1.60E−02	8.00E−03	3.20E−02
TMAH	1.76E−04	8.81E−05	3.52E−04
TDEAH	0.00E+00	0.00E+00	0.00E+00
TDMAS	0.00E+00	0.00E+00	0.00E+00
DEA	0.00E+00	0.00E+00	0.00E+00
PDMAT	0.00E+00	0.00E+00	0.00E+00
Flammable			
SiH$_4$	4.40E−04	2.20E−04	8.80E−04
H$_2$	5.07E−03	2.53E−03	1.01E−02
DCS	0.00E+00	0.00E+00	0.00E+00
B$_2$H$_6$	0.00E+00	0.00E+00	0.00E+00
C$_4$F$_8$	3.30E−04	1.65E−04	6.59E−04
GeH$_4$	0.00E+00	0.00E+00	0.00E+00
C$_2$H$_2$			

(continued)

Table A.33   (continued)

45 nm	Emissions to air		
kg/wafer	Expected value	Low	High
Other chemicals			
Ar	1.91E+00	9.55E−01	3.82E+00
N_2	1.52E+00	7.60E−01	3.04E+00
He	1.49E−01	7.44E−02	2.98E−01
O_2	5.87E−02	2.93E−02	1.17E−01
IPA	3.65E−02	1.83E−02	7.30E−02
NH_3	1.49E−01	7.44E−02	2.98E−01
Ti	9.66E−08	4.83E−08	1.93E−07
Cu	5.22E−05	2.61E−05	1.04E−04
Sn	4.33E−03	2.17E−03	8.67E−03

Table A.34   CMOS Manufacturing emissions per wafer, 32 nm node

32 nm	Emissions to air		
g/wafer	Expected value	Low	High
GWG			
CF_4	2.42E−03	1.21E−03	4.83E−03
CHF_3	1.72E−05	8.59E−06	3.43E−05
C_2F_6	4.40E−06	2.20E−06	8.80E−06
CH_4	4.12E−03	2.06E−03	8.24E−03
CO_2	1.75E+00	8.75E−01	3.50E+00
NF_3	4.27E−04	2.14E−04	8.54E−04
C_4F_6	1.41E−05	7.05E−06	2.82E−05
N_2O	0.00E+00	0.00E+00	0.00E+00
Hazardous			
F_2	1.23E−03	6.16E−04	2.46E−03
HF (gas)	9.23E−03	4.61E−03	1.85E−02
SiF_4	8.58E−05	4.29E−05	1.72E−04
C_4F_8	1.50E−06	7.50E−07	3.00E−06
C_2F_4	3.54E−06	1.77E−06	7.08E−06
COF_2	1.63E−05	8.17E−06	3.27E−05
HCl (gas)	1.09E−07	5.44E−08	2.18E−07
Cl_2	5.55E−05	2.78E−05	1.11E−04
$SiCl_4$	7.48E−07	3.74E−07	1.50E−06
PH_3	0.00E+00	0.00E+00	0.00E+00
BTBAS	4.21E−06	2.10E−06	8.42E−06
AsH_3	0.00E+00	0.00E+00	0.00E+00
BCl_3	1.52E−05	7.62E−06	3.05E−05
Pb	0.00E+00	0.00E+00	0.00E+00
WF_6	1.79E−03	8.97E−04	3.59E−03

(continued)

Table A.34 (continued)

32 nm	Emissions to air		
g/wafer	Expected value	Low	High
HBr	0.00E+00	0.00E+00	0.00E+00
Br$_2$	0.00E+00	0.00E+00	0.00E+00
HMDS	0.00E+00	0.00E+00	0.00E+00
OMCTS	0.00E+00	0.00E+00	0.00E+00
Criteria			
CO	8.11E−01	4.06E−01	1.62E+00
NO	8.66E−03	4.33E−03	1.73E−02
NO$_2$	8.90E−03	4.45E−03	1.78E−02
SO$_2$	0.00E+00	0.00E+00	0.00E+00
N$_2$O	2.20E−04	1.10E−04	4.40E−04
O$_3$	0.00E+00	0.00E+00	0.00E+00

Table A.35 CMOS Manufacturing emissions per wafer, 32 nm node

32 nm	Emissions to air		
g/wafer	Expected value	Low	High
VOC			
DMA	2.13E−06	1.06E−06	4.26E−06
MMA	4.90E−05	2.45E−05	9.80E−05
TDMAT	0.00E+00	0.00E+00	0.00E+00
TMS	1.80E−06	9.00E−07	3.60E−06
TEOS	1.65E−07	8.25E−08	3.30E−07
formaldehyde	2.17E−04	1.09E−04	4.34E−04
m-cresol	1.38E−05	6.88E−06	2.75E−05
p-cresol	1.36E−05	6.82E−06	2.73E−05
PGMEA	2.55E−02	1.28E−02	5.11E−02
n-methyl-2-pyrollidone	4.54E−04	2.27E−04	9.09E−04
PGME	1.77E−02	8.87E−03	3.55E−02
ethyl lactate	2.18E−02	1.09E−02	4.36E−02
TMAH	2.40E−04	1.20E−04	4.80E−04
TDEAH	0.00E+00	0.00E+00	0.00E+00
TDMAS	0.00E+00	0.00E+00	0.00E+00
DEA	0.00E+00	0.00E+00	0.00E+00
PDMAT	0.00E+00	0.00E+00	0.00E+00
Flammable			
SiH$_4$	3.83E−04	1.91E−04	7.65E−04
H$_2$	2.94E−03	1.47E−03	5.89E−03
DCS	0.00E+00	0.00E+00	0.00E+00
B$_2$H$_6$	0.00E+00	0.00E+00	0.00E+00
C$_4$F$_8$	4.40E−04	2.20E−04	8.80E−04
GeH$_4$	0.00E+00	0.00E+00	0.00E+00
C$_2$H$_2$			

(continued)

Table A.35 (continued)

32 nm	Emissions to air		
g/wafer	Expected value	Low	High
Other chemicals			
Ar	2.32E+00	1.16E+00	4.63E+00
N_2	1.46E+00	7.31E−01	2.92E+00
He	1.43E−01	7.14E−02	2.86E−01
O_2	4.34E−02	2.17E−02	8.68E−02
IPA	4.93E−02	2.46E−02	9.85E−02
NH_3	1.43E−01	7.14E−02	2.86E−01
Ti	9.47E−13	4.74E−13	1.89E−12
Cu	4.01E−06	2.01E−06	8.02E−06
Sn	3.04E−04	1.52E−04	6.08E−04

A.3   CMOS Logic Impact Results

Table A.36 Life cycle impacts per die, 32 nm node

	Photochemical smog kg NOx			Acidification mol H+			Ecotoxicity kg 2,4-D		
	Expected value	Lower bound	Upper bound	Expected value	Lower bound	Upper bound	Expected value	Lower bound	Upper bound
Transportation	6.21E−04	N/A	N/A	1.98E−02	N/A	N/A	0.00E+00	N/A	N/A
Electricity (Si, fab. and use)	0.54	0.31	2.19	31.2	17.6	125.9	2.67	1.51	10.8
Fab. direct emissions and EOL	3.52E−04	1.76E−04	7.04E−04	0.28	0.14	0.55	6.58E−04	5.92E−04	7.24E−04
Infrastructure	1.04E−02	5.19E−03	1.56E−02	0.54	0.27	0.81	6.94E−05	3.47E−05	1.04E−04
Total	0.56	0.31	2.2	32	18	127	2.7	1.51	10.8

	Human Health Cancer kg C6H6			Human health criteria Total DALYs			Human Health noncancer kg C7H7		
	Expected value	Lower bound	Upper bound	Expected value	Lower bound	Upper bound	Expected value	Lower bound	Upper bound
Transportation	0.00E+00	N/A	N/A	1.09E−06	N/A	N/A	1.39E−04	N/A	N/A
Electricity (Si, fab. and use)	0.00E+00	0.00E+00	0.00E+00	4.72E−03	2.63E−03	1.84E−02	214	121	865
Fab. direct emissions and EOL	2.65E−05	2.33E−05	2.96E−05	1.45E−07	7.23E−08	2.89E−07	1.4	1.3	1.6
Infrastructure	1.03E−04	5.13E−05	1.54E−04	1.84E−04	9.20E−05	2.76E−04	4.3	2.2	6.5
Total	1.29E−04	7.46E−05	1.83E−04	4.90E−03	2.72E−03	1.87E−02	220	124	873

	Eutrophication, to air kg N			Eutrophication, to water kg N		
	Expected value	Lower bound	Upper bound	Expected value	Lower bound	Upper bound
Transportation	2.19E−05	N/A	N/A	N/A	N/A	N/A
Electricity (Si, fab. and use)	1.94E−02	1.09E−02	7.83E−02	1.38E−02	7.04E−03	2.73E−02
Fab. direct emissions and EOL	4.05E−06	2.02E−06	8.09E−06	N/A	N/A	N/A
Infrastructure	3.53E−04	1.76E−04	5.29E−04	N/A	N/A	N/A
Total	1.98E−02	1.11E−02	7.88E−02	1.38E−02	7.04E−03	2.73E−02

Table A.37 Life cycle impacts per die, 45 nm node

	Photochemical smog kg NOx			Acidification mol H+			Ecotoxicity kg 2,4-D		
	Expected value	Lower bound	Upper bound	Expected value	Lower bound	Upper bound	Expected value	Lower bound	Upper bound
Transportation	7.69E−04	N/A	N/A	2.45E−02	N/A	N/A	0.00E+00	N/A	N/A
Electricity (Si, fab. and use)	0.38	0.22	1.55	22.0	12.4	88.8	1.89	1.06	7.6
Fab. direct emissions and EOL	2.67E−04	1.33E−04	5.33E−04	0.18	0.09	0.36	5.98E−04	4.95E−04	7.54E−04
Infrastructure	9.68E−03	4.84E−03	1.45E−02	0.51	0.25	0.76	6.48E−05	3.24E−05	9.72E−05
Total	0.39	0.22	1.6	23	13	90	1.9	1.06	7.6

	Human Health Cancer kg C6H6			Human health criteria total DALYs			Human Health noncancer kg C7H7		
	Expected value	Lower bound	Upper bound	Expected value	Lower bound	Upper bound	Expected value	Lower bound	Upper bound
Transportation	0.00E+00	N/A	N/A	1.36E−06	N/A	N/A	1.72E−04	N/A	N/A
Electricity (Si, fab. and use)	0.00E+00	0.00E+00	0.00E+00	3.33E−03	1.85E−03	1.28E−02	151	85	610
Fab. direct emissions and EOL	2.88E−05	2.53E−05	3.23E−05	1.13E−07	5.63E−08	2.25E−07	3.1	2.8	3.5
Infrastructure	9.57E−05	4.79E−05	1.44E−04	1.72E−04	8.59E−05	2.58E−04	4.1	2.0	6.1
Total	1.25E−04	7.32E−05	1.76E−04	3.50E−03	1.94E−03	1.31E−02	159	90	620

	Eutrophication, to air kg N			Eutrophication, to water kg N		
	Expected value	Lower bound	Upper bound	Expected value	Lower bound	Upper bound
Transportation	2.71E−05	N/A	N/A	N/A	N/A	N/A
Electricity (Si, fab. and use)	1.37E−02	7.73E−03	5.52E−02	1.47E−02	7.47E−03	2.90E−02
Fab. direct emissions and EOL	2.06E−06	1.03E−06	4.12E−06	N/A	7.47E−03	2.90E−02
Infrastructure	3.29E−04	1.65E−04	4.94E−04	N/A	N/A	N/A
Total	1.41E−02	7.89E−03	5.57E−02	1.47E−02	7.47E−03	2.90E−02

Table A.38 Life cycle impacts per die, 65 nm node

	Photochemical smog kg NO_x			Acidification mol H^+			Ecotoxicity kg 2,4-D		
	Expected value	Lower bound	Upper bound	Expected value	Lower bound	Upper bound	Expected value	Lower bound	Upper bound
Transportation	6.21E−04	N/A	N/A	1.98E−02	N/A	N/A	0.00E+00	N/A	N/A
Electricity (Si, fab. and use)	0.22	0.12	0.86	12.4	7.1	49.2	1.06	0.60	4.2
Fab. emissions and EOL	2.53E−04	1.27E−04	5.07E−04	0.48	0.24	0.96	4.38E−04	3.94E−04	4.82E−04
Infrastructure	6.51E−03	3.26E−03	9.77E−03	0.34	0.17	0.51	4.18E−05	2.09E−05	6.27E−05
Total	0.22	0.13	0.87	13	7.5	51	1.06	0.61	4.2

	Human Health Cancer kg C_6H_6			Human health criteria total DALYs			Health noncancer kg C_7H_7		
	Expected value	Lower bound	Upper bound	Expected value	Lower bound	Upper bound	Expected value	Lower bound	Upper bound
Transportation	0.00E+00	N/A	N/A	1.09E−06	N/A	N/A	1.39E−04	N/A	N/A
Electricity (Si, fab. and use)	0.00E+00	0.00E+00	0.00E+00	1.87E−03	1.05E−03	7.05E−03	85	49	338
Fab. emissions and EOL	2.43E−05	2.14E−05	2.72E−05	1.09E−07	5.45E−08	2.18E−07	2.6	2.3	3.0
Infrastructure	6.18E−05	3.09E−05	9.27E−05	1.16E−04	5.80E−05	1.74E−04	2.6	1.3	3.9
Total	8.61E−05	5.22E−05	1.20E−04	1.99E−03	1.11E−03	7.22E−03	90	52	345

	Eutrophication, to air kg N			Eutrophication, to water kg N		
	Expected value	Lower bound	Upper bound	Expected value	Lower bound	Upper bound
Transportation	2.19E−05	N/A	N/A	N/A	N/A	N/A
Electricity (Si, fab. and use)	7.69E−03	4.40E−03	3.06E−02	1.24E−02	6.30E−03	2.44E−02
Fab. emissions and EOL	1.99E−06	9.96E−07	3.99E−06	N/A	N/A	N/A
Infrastructure	2.22E−04	1.11E−04	3.33E−04	N/A	N/A	N/A
Total	7.94E−03	4.51E−03	3.09E−02	1.24E−02	6.30E−03	2.44E−02

Table A.39 Life cycle impacts per die, 90 nm node

	Photochemical smog kg NO$_x$			Acidification mol H$^+$			Ecotoxicity kg 2,4-D		
	Expected value	Lower bound	Upper bound	Expected value	Lower bound	Upper bound	Expected value	Lower bound	Upper bound
Transportation	6.27E−04	N/A	N/A	2.00E−02	N/A	N/A	0.00E+00	N/A	N/A
Electricity (Si, fab. and use)	0.17	0.10	0.69	10.0	5.8	39.6	0.86	0.49	3.39
Fab. emissions and EOL	1.42E−04	7.06E−05	2.82E−04	0.16	0.08	0.32	4.41E−04	3.96E−04	4.86E−04
Infrastructure	4.81E−03	2.40E−03	7.21E−03	0.25	0.13	0.38	3.09E−05	1.54E−05	4.63E−05
Total	0.18	0.10	0.70	10.5	6.0	40	0.86	0.49	3.4

	Human Health Cancer kg C$_6$H$_6$			Human health criteria total DALYs			Health noncancer kg C$_7$H$_7$		
	Expected value	Lower bound	Upper bound	Expected value	Lower bound	Upper bound	Expected value	Lower bound	Upper bound
Transportation	0.00E+00	N/A	N/A	1.10E−06	N/A	N/A	1.41E−04	N/A	N/A
Electricity (Si, fab. and use)	0.00E+00	0.00E+00	0.00E+00	1.52E−03	8.59E−04	5.70E−03	69	39.6	272
Fab. emissions and EOL	2.92E−05	2.52E−05	3.39E−05	2.29E−08	1.14E−08	4.57E−08	3.0	2.6	3.4
Infrastructure	4.56E−05	2.28E−05	6.84E−05	8.56E−05	4.28E−05	1.28E−04	1.9	1.0	2.9
Total	7.48E−05	4.80E−05	1.02E−04	1.60E−03	9.02E−04	5.83E−03	74	43	278

	Eutrophication, to air kg N			Eutrophication, to water kg N		
	Expected value	Lower bound	Upper bound	Expected value	Lower bound	Upper bound
Transportation	2.21E−05	N/A	N/A	N/A	N/A	N/A
Electricity (Si, fab. and use)	6.23E−03	3.59E−03	2.46E−02	1.41E−02	7.20E−03	2.79E−02
Fab. direct emissions and EOL	5.43E−07	2.72E−07	1.09E−06	N/A	N/A	N/A
Infrastructure	1.64E−04	8.18E−05	2.46E−04	N/A	N/A	N/A
Total	6.42E−03	3.67E−03	2.49E−02	1.41E−02	7.20E−03	2.79E−02

Table A.40 Life cycle impacts per die, 130 nm node

	Photochemical smog kg NOx			Acidification mol H+			Ecotoxicity kg 2,4-D		
	Expected value	Lower bound	Upper bound	Expected value	Lower bound	Upper bound	Expected value	Lower bound	Upper bound
Transportation	6.27E−04	N/A	N/A	2.00E−02	N/A	N/A	0.00E+00	N/A	N/A
Electricity (Si, fab. and use)	0.13	0.074	0.496	7.3	4.3	28.5	0.63	0.36	2.44
Fab. direct emissions and EOL	1.21E−04	6.01E−05	2.40E−04	0.09	0.04	0.18	3.80E−04	3.41E−04	4.18E−04
Infrastructure	2.79E−03	1.39E−03	4.18E−03	0.15	0.07	0.22	1.99E−05	9.96E−06	2.99E−05
Total	0.13	0.08	0.50	7.6	4.4	29	0.63	0.37	2.4

	Human Health Cancer kg C6H6			Human health criteria total DALYs			Health noncancer kg C7H7		
	Expected value	Lower bound	Upper bound	Expected value	Lower bound	Upper bound	Expected value	Lower bound	Upper bound
Transportation	0.00E+00	N/A	N/A	1.10E−06	N/A	N/A	1.41E−04	N/A	N/A
Electricity (Si, fab. and use)	0.00E+00	0.00E+00	0.00E+00	1.11E−03	6.38E−04	4.16E−03	50	29.3	196
Fab. emissions and EOL	1.21E−05	1.02E−05	1.48E−05	2.22E−08	1.11E−08	4.44E−08	2.9	2.5	3.2
Infrastructure	2.94E−05	1.47E−05	4.42E−05	4.91E−05	2.46E−05	7.37E−05	1.2	0.6	1.9
Total	4.16E−05	2.49E−05	5.89E−05	1.16E−03	6.63E−04	4.23E−03	54	32	201

	Eutrophication, to air kg N			Eutrophication, to water kg N		
	Expected value	Lower bound	Upper bound	Expected value	Lower bound	Upper bound
Transportation	2.21E−05	N/A	N/A	N/A	N/A	N/A
Electricity (Si, fab. and uses)	4.55E−03	2.65E−03	1.77E−02	1.38E−02	6.95E−03	2.75E−02
Fab. direct emissions and EOL	5.33E−07	2.67E−07	1.07E−06	N/A	N/A	N/A
Infrastructure	9.47E−05	4.74E−05	1.42E−04	N/A	N/A	N/A
Total	4.66E−03	2.70E−03	1.79E−02	1.38E−02	6.95E−03	2.75E−02

Table A.41 Life cycle impacts per die, 180 nm node

	Photochemical smog kg NO$_x$			Acidification mol H$^+$			Ecotoxicity kg 2,4-D		
	Expected value	Lower bound	Upper bound	Expected value	Lower bound	Upper bound	Expected value	Lower bound	Upper bound
Transportation	6.88E−04	N/A	N/A	2.20E−02	N/A	N/A	0.00E+00	N/A	N/A
Electricity (Si, fab. and use)	0.053	0.033	0.190	3.0	1.9	10.9	0.26	0.16	0.94
Fab. direct emissions and EOL	1.76E−04	8.78E−05	3.51E−04	0.08	0.04	0.15	2.64E−05	2.32E−05	3.03E−05
Infrastructure	4.21E−03	2.10E−03	6.31E−03	0.22	0.11	0.33	2.92E−05	1.46E−05	4.38E−05
Total	0.058	0.035	0.197	3.4	2.0	11.4	0.26	0.16	0.94

	Human Health Cancer kg C$_6$H$_6$			Human health criteria total DALYs			Health noncancer kg C$_7$H$_7$		
	Expected value	Lower bound	Upper bound	Expected value	Lower bound	Upper bound	Expected value	Lower bound	Upper bound
Transportation	0.00E+00	N/A	N/A	1.21E−06	N/A	N/A	1.55E−04	N/A	N/A
Electricity (Si, fab. and use)	0.00E+00	0.00E+00	0.00E+00	4.63E−04	2.76E−04	1.44E−03	21	13.0	75
Fab. emissions and EOL	1.68E−05	1.41E−05	2.03E−05	4.37E−08	2.19E−08	8.74E−08	2.7	2.4	3.1
Infrastructure	4.31E−05	2.16E−05	6.47E−05	7.44E−05	3.72E−05	1.12E−04	1.8	0.9	2.7
Total	5.99E−05	3.56E−05	8.50E−05	5.39E−04	3.14E−04	1.55E−03	25	16.3	81

	Eutrophication, to air kg N			Eutrophication, to water kg N		
	Expected value	Lower bound	Upper bound	Expected value	Lower bound	Upper bound
Transportation	2.43E−05	N/A	N/A	N/A	N/A	N/A
Electricity (Si, fab. and use)	1.89E−03	1.18E−03	6.80E−03	1.92E−02	9.68E−03	3.82E−02
Fab. direct emissions and EOL	1.05E−06	5.25E−07	2.10E−06	N/A	N/A	N/A
Infrastructure	1.43E−04	7.15E−05	2.15E−04	N/A	N/A	N/A
Total	2.06E−03	1.25E−03	7.01E−03	1.92E−02	9.68E−03	3.82E−02

Table A.42 Life cycle impacts per die, 250 nm node

	Photochemical smog kg NO$_x$			Acidification mol H$^+$			Ecotoxicity kg 2,4-D		
	Expected value	Lower bound	Upper bound	Expected value	Lower bound	Upper bound	Expected value	Lower bound	Upper bound
Transportation	6.56E−04	N/A	N/A	2.09E−02	N/A	N/A	0.00E+00	N/A	N/A
Electricity (Si, fab. and use)	0.049	0.032	0.171	2.8	1.8	9.8	0.24	0.16	0.84
Fab. emissions and EOL	1.64E−04	8.17E−05	3.27E−04	0.06	0.03	0.12	3.30E−05	2.90E−05	3.80E−05
Infrastructure	4.83E−03	2.42E−03	7.25E−03	0.25	0.13	0.38	3.35E−05	1.68E−05	5.03E−05
Total	0.055	0.034	0.179	3.2	2.0	10.3	0.24	0.16	0.84

	Human Health Cancer kg C$_6$H$_6$			Human health criteria total DALYs			Health noncancer kg C$_7$H$_7$		
	Expected value	Lower bound	Upper bound	Expected value	Lower bound	Upper bound	Expected value	Lower bound	Upper bound
Transportation	0.00E+00	N/A	N/A	1.16E−06	N/A	N/A	1.47E−04	N/A	N/A
Electricity (Si, fab. and use)	0.00E+00	0.00E+00	0.00E+00	4.33E−04	2.63E−04	1.25E−03	19	13	68
Fab. emissions and EOL	4.77E−05	4.11E−05	5.55E−05	5.58E−10	2.79E−10	1.12E−09	3.4	3.0	3.9
Infrastructure	4.95E−05	2.48E−05	7.43E−05	8.55E−05	4.28E−05	1.28E−04	2.1	1.1	3.2
Total	9.73E−05	6.59E−05	1.30E−04	5.19E−04	3.06E−04	1.38E−03	25	17	75

	Eutrophication, to air kg N			Eutrophication, to water kg N		
	Expected value	Lower bound	Upper bound	Expected value	Lower bound	Upper bound
Transportation	2.31E−05	N/A	N/A	N/A	N/A	N/A
Electricity (Si, fab. and use)	1.76E−03	1.14E−03	6.12E−03	2.31E−02	1.18E−02	4.56E−02
Fab. direct emissions and EOL	2.59E−07	1.30E−07	5.19E−07	N/A	N/A	N/A
Infrastructure	1.64E−04	8.22E−05	2.47E−04	2.31E−02	1.18E−02	4.56E−02
Total	1.95E−03	1.22E−03	6.37E−03	2.31E−02	1.18E−02	4.56E−02

Table A.43 Life cycle impacts per die, 350 nm node

	Photochemical smog kg NO_x			Acidification mol H^+			Ecotoxicity kg 2,4-D		
	Expected value	Lower bound	Upper bound	Expected value	Lower bound	Upper bound	Expected value	Lower bound	Upper bound
Transportation	7.38E−04	N/A	N/A	2.35E−02	N/A	N/A	0.00E+00	N/A	N/A
Electricity (Si, fab. and use)	0.033	0.024	0.092	1.9	1.4	5.3	0.16	0.12	0.45
Fab. emissions and EOL	2.54E−04	1.27E−04	5.06E−04	0.09	0.05	0.18	5.23E−05	4.58E−05	6.04E−05
Infrastructure	5.79E−03	2.90E−03	8.69E−03	0.30	0.15	0.45	3.72E−05	1.86E−05	5.58E−05
Total	0.039	0.027	0.101	2.3	1.6	5.9	0.16	0.12	0.45

	Human Health Cancer kg C_6H_6			Human health criteria total DALYs			Health noncancer kg C_7H_7		
	Expected value	Lower bound	Upper bound	Expected value	Lower bound	Upper bound	Expected value	Lower bound	Upper bound
Transportation	0.00E+00	N/A	N/A	1.30E−06	N/A	N/A	1.66E−04	N/A	N/A
Electricity (Si, fab. and use)	0.00E+00	0.00E+00	0.00E+00	2.88E−04	1.91E−04	5.60E−04	13	9.5	36
Fab. emissions and EOL	7.18E−05	6.17E−05	8.40E−05	9.59E−10	4.80E−10	1.92E−09	5.2	4.5	6.0
Infrastructure	5.49E−05	2.75E−05	8.24E−05	1.03E−04	5.15E−05	1.55E−04	2.3	1.2	3.5
Total	1.27E−04	8.91E−05	1.66E−04	3.92E−04	2.42E−04	7.14E−04	20	15	46

	Eutrophication, to air kg N			Eutrophication, to water kg N		
	Expected value	Lower bound	Upper bound	Expected value	Lower bound	Upper bound
Transportation	2.60E−05	N/A	N/A	N/A	N/A	N/A
Electricity (Si, fab. and use)	1.16E−03	8.59E−04	3.27E−03	3.44E−02	1.75E−02	6.80E−02
Fab. direct emissions and EOL	3.88E−07	1.94E−07	7.76E−07	N/A	N/A	N/A
Infrastructure	1.97E−04	9.85E−05	2.96E−04	N/A	N/A	N/A
Total	1.39E−03	9.57E−04	3.56E−03	3.44E−02	1.75E−02	6.80E−02

Appendix B
Flash Memory Life Cycle Inventory and Impact Results

B.1 Flash Memory: Manufacturing Inventory Data

Table B.1 Flash memory manufacturing inventory per wafer, 150 nm node

150 nm		150 nm	
Inputs	g/wafer	Inputs	g/wafer
GWG		*Flammable*	
CF_4	1.63E+01	SiH_4	1.32E+01
CHF_3	3.52E+00	H_2	9.58E+01
C_2F_6	0.00E+00	DCS	0.00E+00
CH_4	5.32E+02	B_2H_6	0.00E+00
CO_2	0.00E+00	C_4F_8	0.00E+00
NF_3	1.57E+02	GeH_4	0.00E+00
C_4F_6	0.00E+00	C_2H_2	0.00E+00
N_2O	0.00E+00	C_2H_4	0.00E+00
Hazardous			
F_2	0.00E+00	*Other chemicals*	
HF (gas)	0.00E+00	O_2	4.20E+02
SiF_4	1.15E−01	Ar	6.62E+02
C_4F_8	1.06E+00	N_2	7.71E+02
C_2F_4	0.00E+00	CMP polishing solution	0.00E+00
COF_2	0.00E+00	H_2O_2	1.30E+04
HCl (gas)	0.00E+00	abrasive	0.00E+00
Cl_2	2.79E+00	silica slurry	0.00E+00
$SiCl_4$	0.00E+00	benzotriazole	0.00E+00
PH_3	2.84E+00	CMP surfactants	1.23E+02

(continued)

S.B. Boyd, *Life-Cycle Assessment of Semiconductors*, DOI 10.1007/978-1-4419-9988-7,
© Springer Science+Business Media, LLC 2012

Table B.1 (continued)

150 nm		150 nm	
Inputs	g/wafer	Inputs	g/wafer
BTBAS	2.05E+00	citric acid	6.84E+01
AsH$_3$	2.62E−01	tungsten CMP slurry	5.92E+03
BCl$_3$	1.36E+00	Al	1.22E−01
Pb	7.19E−02	As	0.00E+00
WF$_6$	3.48E+01	Cu	8.35E−04
HBr	8.48E+00	Pt	0.00E+00
Br$_2$	0.00E+00	Ti	6.05E−04
HMDS	2.56E+02	He	4.78E−01
OMCTS	0.00E+00	NH$_3$	2.16E+01

Table B.2 Flash memory manufacturing inventory per wafer, 150 nm node

150 nm		150 nm	
Inputs	g/wafer	Inputs	g/wafer
Criteria		*Other chemicals*	
CO	0.00E+00	CuSO$_4$	0.00E+00
NO	0.00E+00	ArH	0.00E+00
NO$_2$	0.00E+00	100% HF liquid	1.43E+02
SO$_2$	0.00E+00	H$_2$SO$_4$	1.12E+05
N$_2$O	7.40E+00	HCl	4.23E+03
O$_3$	0.00E+00	NH$_4$OH	6.01E+03
		IPA	8.33E+02
VOC		oxide CMP slurry, chemicals	1.14E+04
DMA	0.00E+00	BF$_3$	2.05E−02
MMA	0.00E+00	Cr	0.00E+00
TDMAT	6.46E−01	Au	0.00E+00
TMS	0.00E+00	Sn	1.99E−02
TEOS	0.00E+00	Co	0.00E+00
formaldehyde	3.05E+01	Ni	0.00E+00
m-cresol	1.93E+01	polyamide laminate	2.65E+00
p-cresol	1.91E+01	laminate solvent	0.00E+00
PGMEA	2.46E+02	H$_3$PO$_4$	0.00E+00
n-methyl-2-pyrollidone	1.19E+01	CuCl$_2$	0.00E+00
PGME	3.23E+02	W	2.71E−03
ethyl lactate	3.09E+02	Ta	0.00E+00
TMAH	3.21E+01	Cu ECP additive X	0.00E+00
TDEAH	0.00E+00	Cu ECP Additive Y	0.00E+00
TDMAS	0.00E+00	Na$_2$B$_4$O$_7$	1.39E+01
DEA	0.00E+00		
PDMAT	0.00E+00		

Table B.3 Flash memory manufacturing inventory per wafer, 120 nm node

120 nm		120 nm	
Inputs	g/wafer	Inputs	g/wafer
GWG		*Flammable*	
CF_4	5.88E+00	SiH_4	1.24E+01
CHF_3	3.31E+00	H_2	9.01E+01
C_2F_6	0.00E+00	DCS	0.00E+00
CH_4	5.00E+02	B_2H_6	0.00E+00
CO_2	0.00E+00	C_4F_8	0.00E+00
NF_3	1.47E+02	GeH_4	0.00E+00
C_4F_6	0.00E+00	C_2H_2	0.00E+00
N_2O	0.00E+00	C_2H_4	0.00E+00
Hazardous			
F_2	0.00E+00	*Other chemicals*	
HF (gas)	0.00E+00	O_2	3.00E+02
SiF_4	1.08E−01	Ar	6.23E+02
C_4F_8	9.94E−01	N_2	7.24E+02
C_2F_4	0.00E+00	CMP polishing solution	0.00E+00
COF_2	0.00E+00	H_2O_2	1.23E+04
HCl (gas)	0.00E+00	abrasive	0.00E+00
Cl_2	2.63E+00	silica slurry	0.00E+00
$SiCl_4$	0.00E+00	benzotriazole	0.00E+00
PH_3	2.67E+00	CMP surfactants	1.16E+02
BTBAS	1.93E+00	citric acid	6.43E+01
AsH_3	2.75E−01	tungsten CMP slurry	5.56E+03
BCl_3	1.28E+00	Al	1.15E−01
Pb	6.76E−02	As	0.00E+00
WF_6	3.27E+01	Cu	7.85E−04
HBr	7.97E+00	Pt	0.00E+00
Br_2	0.00E+00	Ti	5.69E−04
HMDS	2.59E+02	He	4.50E−01
OMCTS	0.00E+00	NH_3	2.03E+01

Table B.4 Flash memory manufacturing inventory per wafer, 120 nm node

120 nm		120 nm	
Inputs	g/wafer	Inputs	g/wafer
Criteria		*Other chemicals*	
CO	0.00E+00	$CuSO_4$	0.00E+00
NO	0.00E+00	ArH	0.00E+00
NO_2	0.00E+00	100% HF liquid	1.34E+02
SO_2	0.00E+00	H_2SO_4	1.06E+05
N_2O	6.96E+00	HCl	3.98E+03
O_3	0.00E+00	NH_4OH	5.64E+03
		IPA	7.83E+02

(continued)

Table B.4 (continued)

120 nm		120 nm	
Inputs	g/wafer	Inputs	g/wafer
VOC		oxide CMP slurry, chemicals	1.07E+04
DMA	0.00E+00	BF_3	1.92E−02
MMA	0.00E+00	Cr	0.00E+00
TDMAT	6.07E−01	Au	0.00E+00
TMS	0.00E+00	Sn	1.87E−02
TEOS	0.00E+00	Co	0.00E+00
formaldehyde	3.08E+01	Ni	0.00E+00
m-cresol	1.95E+01	polyamide laminate	2.49E+00
p-cresol	1.94E+01	laminate solvent	0.00E+00
PGMEA	2.49E+02	H_3PO_4	0.00E+00
n-methyl-2-pyrollidone	1.12E+01	$CuCl_2$	0.00E+00
PGME	3.27E+02	W	2.55E−03
ethyl lactate	3.13E+02	Ta	0.00E+00
TMAH	3.25E+01	Cu ECP additive X	0.00E+00
TDEAH	0.00E+00	Cu ECP Additive Y	0.00E+00
TDMAS	0.00E+00	$Na_2B_4O_7$	1.41E+01
DEA	0.00E+00		
PDMAT	0.00E+00		

Table B.5 Flash memory manufacturing inventory per wafer, 90 nm node

90 nm		90 nm	
Inputs	g/wafer	Inputs	g/wafer
GWG		*Flammable*	
CF_4	5.88E+00	SiH_4	1.24E+01
CHF_3	3.31E+00	H_2	9.01E+01
C_2F_6	0.00E+00	DCS	0.00E+00
CH_4	5.32E+02	B_2H_6	0.00E+00
CO_2	0.00E+00	C_4F_8	0.00E+00
NF_3	1.47E+02	GeH_4	0.00E+00
C_4F_6	0.00E+00	C_2H_2	0.00E+00
N_2O	0.00E+00	C_2H_4	0.00E+00
Hazardous			
F_2	0.00E+00	*Other chemicals*	
HF (gas)	0.00E+00	O_2	3.00E+02
SiF_4	1.08E−01	Ar	6.23E+02
C_4F_8	9.94E−01	N_2	7.24E+02
C_2F_4	0.00E+00	CMP polishing solution	0.00E+00
COF_2	0.00E+00	H_2O_2	1.23E+04
HCl (gas)	0.00E+00	abrasive	0.00E+00
Cl_2	2.63E+00	silica slurry	0.00E+00

(continued)

Table B.5 (continued)

90 nm		90 nm	
Inputs	g/wafer	Inputs	g/wafer
SiCl$_4$	0.00E+00	benzotriazole	0.00E+00
PH$_3$	2.67E+00	CMP surfactants	1.16E+02
BTBAS	1.93E+00	citric acid	6.43E+01
AsH$_3$	2.75E−01	tungsten CMP slurry	5.56E+03
BCl$_3$	1.28E+00	Al	1.15E−01
Pb	0.00E+00	As	0.00E+00
WF$_6$	3.27E+01	Cu	1.57E−03
HBr	7.97E+00	Pt	0.00E+00
Br$_2$	0.00E+00	Ti	5.69E−04
HMDS	2.59E+02	He	4.50E−01
OMCTS	0.00E+00	NH$_3$	2.03E+01

Table B.6 Flash memory manufacturing inventory per wafer, 90 nm node

90 nm		90 nm	
Inputs	g/wafer	Inputs	g/wafer
Criteria		*Other chemicals*	
CO	0.00E+00	CuSO$_4$	0.00E+00
NO	0.00E+00	ArH	0.00E+00
NO$_2$	0.00E+00	100% HF liquid	1.34E+02
SO$_2$	0.00E+00	H$_2$SO$_4$	1.06E+05
N$_2$O	6.96E+00	HCl	3.98E+03
O$_3$	0.00E+00	NH$_4$OH	5.64E+03
		IPA	7.83E+02
VOC		oxide CMP slurry, chemicals	1.07E+04
DMA	0.00E+00	BF$_3$	1.92E−02
MMA	0.00E+00	Cr	0.00E+00
TDMAT	6.07E−01	Au	0.00E+00
TMS	0.00E+00	Sn	5.95E−02
TEOS	0.00E+00	Co	0.00E+00
formaldehyde	3.08E+01	Ni	0.00E+00
m-cresol	1.95E+01	polyamide laminate	2.49E+00
p-cresol	1.94E+01	laminate solvent	0.00E+00
PGMEA	2.49E+02	H$_3$PO$_4$	0.00E+00
n-methyl-2-pyrollidone	1.12E+01	CuCl$_2$	0.00E+00
PGME	3.27E+02	W	2.55E−03
ethyl lactate	3.13E+02	Ta	0.00E+00
TMAH	3.25E+01	Cu ECP additive X	0.00E+00
TDEAH	0.00E+00	Cu ECP Additive Y	0.00E+00
TDMAS	0.00E+00	Na$_2$B$_4$O$_7$	1.41E+01
DEA	0.00E+00		
PDMAT	0.00E+00		

Table B.7 Flash memory manufacturing inventory per wafer, 65 nm node

65 nm		65 nm	
Inputs	g/wafer	Inputs	g/wafer
GWG		*Flammable*	
CF_4	5.86E+00	SiH_4	1.24E+01
CHF_3	3.36E+00	H_2	1.71E+02
C_2F_6	0.00E+00	DCS	0.00E+00
CH_4	5.57E+02	B_2H_6	0.00E+00
CO_2	0.00E+00	C_4F_8	0.00E+00
NF_3	1.57E+02	GeH_4	0.00E+00
C_4F_6	0.00E+00	C_2H_2	0.00E+00
N_2O	0.00E+00	C_2H_4	0.00E+00
Hazardous			
F_2	0.00E+00	*Other chemicals*	
HF (gas)	0.00E+00	O_2	2.96E+02
SiF_4	1.02E−01	Ar	7.64E+02
C_4F_8	1.69E+00	N_2	8.90E+02
C_2F_4	0.00E+00	CMP polishing solution	0.00E+00
COF_2	0.00E+00	H_2O_2	1.22E+04
HCl (gas)	0.00E+00	abrasive	0.00E+00
Cl_2	3.88E+00	silica slurry	0.00E+00
$SiCl_4$	0.00E+00	benzotriazole	0.00E+00
PH_3	2.52E+00	CMP surfactants	1.64E+02
BTBAS	1.82E+00	citric acid	9.09E+01
AsH_3	2.59E−01	tungsten CMP slurry	7.87E+03
BCl_3	2.42E+00	Al	4.34E−01
Pb	0.00E+00	As	0.00E+00
WF_6	6.17E+01	Cu	1.48E−03
HBr	1.00E+01	Pt	0.00E+00
Br_2	0.00E+00	Ti	3.07E−02
HMDS	2.62E+02	He	8.48E−01
OMCTS	0.00E+00	NH_3	1.91E+01

Table B.8 Flash memory manufacturing inventory per wafer, 65 nm node

65 nm		65 nm	
Inputs	g/wafer	Inputs	g/wafer
Criteria		*Other chemicals*	
CO	0.00E+00	$CuSO_4$	0.00E+00
NO	0.00E+00	ArH	0.00E+00
NO_2	0.00E+00	100% HF liquid	1.14E+02
SO_2	0.00E+00	H_2SO_4	9.97E+04
N_2O	6.56E+00	HCl	4.11E+03
O_3	0.00E+00	NH_4OH	6.11E+03
		IPA	7.38E+02

(continued)

Table B.8 (continued)

65 nm		65 nm	
Inputs	g/wafer	Inputs	g/wafer
VOC		oxide CMP slurry, chemicals	1.51E+04
DMA	0.00E+00	BF_3	1.81E−02
MMA	0.00E+00	Cr	0.00E+00
TDMAT	1.15E+00	Au	0.00E+00
TMS	0.00E+00	Sn	5.61E−02
TEOS	0.00E+00	Co	0.00E+00
formaldehyde	3.11E+01	Ni	0.00E+00
m-cresol	1.98E+01	polyamide laminate	2.35E+00
p-cresol	1.96E+01	laminate solvent	0.00E+00
PGMEA	2.52E+02	H_3PO_4	0.00E+00
n-methyl-2-pyrollidone	1.05E+01	$CuCl_2$	0.00E+00
PGME	3.31E+02	W	2.40E−03
ethyl lactate	3.16E+02	Ta	0.00E+00
TMAH	3.30E+01	Cu ECP additive X	0.00E+00
TDEAH	0.00E+00	Cu ECP Additive Y	0.00E+00
TDMAS	0.00E+00	$Na_2B_4O_7$	1.43E+01
DEA	0.00E+00		
PDMAT	0.00E+00		

Table B.9 Flash memory manufacturing inventory per wafer, 45 nm node

45 nm		45 nm	
Inputs	g/wafer	Inputs	g/wafer
GWG		*Flammable*	
CF_4	2.19E+01	SiH_4	1.12E+01
CHF_3	3.63E+00	H_2	1.28E+02
C_2F_6	0.00E+00	DCS	0.00E+00
CH_4	5.49E+02	B_2H_6	0.00E+00
CO_2	0.00E+00	C_4F_8	0.00E+00
NF_3	1.87E+02	GeH_4	0.00E+00
C_4F_6	1.36E−01	C_2H_2	0.00E+00
N_2O	0.00E+00	C_2H_4	0.00E+00
Hazardous			
F_2	0.00E+00	*Other chemicals*	
HF (gas)	0.00E+00	O_2	8.70E+02
SiF_4	3.58E−01	Ar	1.70E+03
C_4F_8	2.06E+00	N_2	6.71E+02
C_2F_4	0.00E+00	CMP polishing solution	7.87E+03
COF_2	0.00E+00	H_2O_2	1.34E+04
HCl (gas)	0.00E+00	abrasive	2.49E+02
Cl_2	3.01E+00	silica slurry	2.49E+02
$SiCl_4$	0.00E+00	benzotriazole	1.02E+01

(continued)

Table B.9 (continued)

45 nm		45 nm	
Inputs	g/wafer	Inputs	g/wafer
PH_3	2.62E+00	CMP surfactants	2.73E+02
BTBAS	0.00E+00	citric acid	6.18E+01
AsH_3	2.59E−01	tungsten CMP slurry	5.24E+03
BCl_3	1.21E+00	Al	1.08E−01
Pb	0.00E+00	As	0.00E+00
WF_6	4.63E+01	Cu	1.48E−03
HBr	1.00E+01	Pt	0.00E+00
Br_2	0.00E+00	Ti	5.37E−04
HMDS	3.58E+02	He	4.24E−01
OMCTS	0.00E+00	NH_3	1.51E+00

Table B.10 Flash memory manufacturing inventory per wafer, 45 nm node

45 nm		45 nm	
Inputs	g/wafer	Inputs	g/wafer
Criteria		*Other chemicals*	
CO	0.00E+00	$CuSO_4$	7.08E+00
NO	0.00E+00	ArH	0.00E+00
NO_2	0.00E+00	100% HF liquid	1.15E+02
SO_2	0.00E+00	H_2SO_4	1.09E+05
N_2O	1.21E+01	HCl	4.47E+03
O_3	0.00E+00	NH_4OH	6.33E+03
		IPA	7.38E+02
VOC		oxide CMP slurry, chemicals	1.51E+04
DMA	0.00E+00	BF_3	1.81E−02
MMA	0.00E+00	Cr	0.00E+00
TDMAT	5.73E−01	Au	0.00E+00
TMS	0.00E+00	Sn	5.61E−02
TEOS	0.00E+00	Co	0.00E+00
formaldehyde	4.26E+01	Ni	2.98E−02
m-cresol	2.70E+01	polyamide laminate	2.35E+00
p-cresol	2.67E+01	laminate solvent	0.00E+00
PGMEA	3.44E+02	H_3PO_4	0.00E+00
n-methyl-2-pyrollidone	1.05E+01	$CuCl_2$	4.06E−03
PGME	4.52E+02	W	2.40E−03
ethyl lactate	4.32E+02	Ta	0.00E+00
TMAH	4.55E+01	Cu ECP additive X	2.06E+00
TDEAH	0.00E+00	Cu ECP Additive Y	1.02E+00
TDMAS	0.00E+00	$Na_2B_4O_7$	1.97E+01
DEA	0.00E+00		
PDMAT	4.01E+00		

B.2   Flash Memory: Manufacturing Emissions Data

Table B.11   Flash memory manufacturing emissions per wafer, 150 nm node

150 nm	Emissions to air		
kg/wafer	Expected value	Low	High
GWG			
CF_4	5.86E−04	8.16E+00	3.27E+01
CHF_3	2.53E−05	1.76E+00	7.03E+00
C_2F_6	0.00E+00	0.00E+00	0.00E+00
CH_4	2.14E−04	2.66E+02	1.06E+03
CO_2	1.46E+00	0.00E+00	0.00E+00
NF_3	3.85E−04	7.84E+01	3.13E+02
C_4F_6	0.00E+00	0.00E+00	0.00E+00
N_2O	0.00E+00	0.00E+00	0.00E+00
Hazardous			
F_2	4.18E−03	0.00E+00	0.00E+00
HF (gas)	2.79E−03	0.00E+00	0.00E+00
SiF_4	5.14E−05	5.77E−02	2.31E−01
C_4F_8	9.62E−07	5.29E−01	2.12E+00
C_2F_4	1.13E−05	0.00E+00	0.00E+00
COF_2	5.34E−05	0.00E+00	0.00E+00
HCl (gas)	1.39E−07	0.00E+00	0.00E+00
Cl_2	6.45E−05	1.40E+00	5.59E+00
$SiCl_4$	9.59E−07	0.00E+00	0.00E+00
PH_3	0.00E+00	1.42E+00	5.67E+00
bis tertiary-butylamino silane	1.62E−05	1.03E+00	4.11E+00
AsH_3	4.08E−06	1.31E−01	5.24E−01
BCl_3	1.95E−05	6.81E−01	2.73E+00
Pb	6.47E−05	3.60E−02	1.44E−01
WF_6	4.60E−03	1.74E+01	6.96E+01
HBr	0.00E+00	4.24E+00	1.70E+01
Br_2	0.00E+00	0.00E+00	0.00E+00
HMDS	0.00E+00	1.28E+02	5.12E+02
OMCTS	0.00E+00	0.00E+00	0.00E+00
Criteria			
CO	2.82E−01	0.00E+00	0.00E+00
NO	7.23E−03	0.00E+00	0.00E+00
NO_2	3.56E−03	0.00E+00	0.00E+00
SO_2	0.00E+00	0.00E+00	0.00E+00
N_2O	2.82E−04	3.70E+00	1.48E+01
O_3	0.00E+00	0.00E+00	0.00E+00

Table B.12 Flash memory manufacturing emissions per wafer, 150 nm node

150 nm	Emissions to air		
kg/wafer	Expected value	Low	High
VOC			
DMA	1.37E−06	0.00E+00	0.00E+00
MMA	3.14E−05	0.00E+00	0.00E+00
TDMAT	0.00E+00	3.23E−01	1.29E+00
TMS	0.00E+00	0.00E+00	0.00E+00
TEOS	0.00E+00	0.00E+00	0.00E+00
formaldehyde (CH_2O)	1.40E−04	1.52E+01	6.09E+01
m-cresol	9.19E−06	9.66E+00	3.86E+01
p-cresol	4.83E−04	9.56E+00	3.83E+01
PGMEA	1.64E−02	1.23E+02	4.93E+02
n-methyl-2-pyrrollidone	6.82E−04	5.94E+00	2.38E+01
PGME	1.20E−02	1.62E+02	6.47E+02
ethyl lactate	1.40E−02	1.55E+02	6.18E+02
TMAH	6.46E−04	1.60E+01	6.41E+01
TDEAH	0.00E+00	0.00E+00	0.00E+00
TDMAS	0.00E+00	0.00E+00	0.00E+00
DEA	0.00E+00	0.00E+00	0.00E+00
PDMAT	0.00E+00	0.00E+00	0.00E+00
Flammable			
SiH_4			
H_2	9.06E−04	6.59E+00	2.63E+01
DCS	4.01E−03	4.79E+01	1.92E+02
B_2H_6	0.00E+00	0.00E+00	0.00E+00
C_4F_8	0.00E+00	0.00E+00	0.00E+00
GeH_4	1.67E−04	0.00E+00	0.00E+00
C_2H_2	0.00E+00	0.00E+00	0.00E+00
C_2H_4			
Other chemicals			
Ar	6.54E−01	3.31E+02	1.32E+03
N_2	7.21E−01	3.85E+02	1.54E+03
He	4.78E−04	2.39E−01	9.57E−01
O_2	9.82E−02	2.10E+02	8.41E+02
IPA	3.38E−02	4.16E+02	1.67E+03
NH_3	4.78E−04	2.39E−01	9.57E−01
Ti	5.45E−07	3.03E−04	1.21E−03
Cu	7.52E−07	4.18E−04	1.67E−03
Sn	1.79E−05	9.94E−03	3.97E−02

Table B.13 Flash memory manufacturing emissions per wafer, 120 nm node

120 nm	Emissions to air		
kg/wafer	Expected value	Low	High
GWG			
CF_4	1.30E−04	2.94E+00	1.18E+01
CHF_3	2.37E−05	1.65E+00	6.61E+00
C_2F_6	0.00E+00	0.00E+00	0.00E+00
CH_4	2.01E−04	2.50E+02	9.99E+02
CO_2	1.37E+00	0.00E+00	0.00E+00
NF_3	3.62E−04	7.36E+01	2.95E+02
C_4F_6	0.00E+00	0.00E+00	0.00E+00
N_2O	0.00E+00	0.00E+00	0.00E+00
Hazardous			
F_2	3.93E−03	0.00E+00	0.00E+00
HF (gas)	6.04E−04	0.00E+00	0.00E+00
SiF_4	3.67E−05	5.42E−02	2.17E−01
C_4F_8	9.04E−07	4.97E−01	1.99E+00
C_2F_4	1.07E−05	0.00E+00	0.00E+00
COF_2	5.02E−05	0.00E+00	0.00E+00
HCl (gas)	1.31E−07	0.00E+00	0.00E+00
Cl_2	6.07E−05	1.31E+00	5.25E+00
$SiCl_4$	9.01E−07	0.00E+00	0.00E+00
PH_3	0.00E+00	1.33E+00	5.34E+00
bis tertiary-butylamino silane	1.52E−05	9.65E−01	3.86E+00
AsH_3	4.29E−06	1.38E−01	5.50E−01
BCl_3	1.84E−05	6.40E−01	2.56E+00
Pb	6.08E−05	3.38E−02	1.35E−01
WF_6	4.32E−03	1.64E+01	6.54E+01
HBr	0.00E+00	3.99E+00	1.59E+01
Br_2	0.00E+00	0.00E+00	0.00E+00
HMDS	0.00E+00	1.30E+02	5.18E+02
OMCTS	0.00E+00	0.00E+00	0.00E+00
Criteria			
CO	2.18E−02	0.00E+00	0.00E+00
NO	4.44E−03	0.00E+00	0.00E+00
NO_2	4.52E−04	0.00E+00	0.00E+00
SO_2	0.00E+00	0.00E+00	0.00E+00
N_2O	2.65E−04	3.48E+00	1.39E+01
O_3	0.00E+00	0.00E+00	0.00E+00

Table B.14 Flash memory manufacturing emissions per wafer, 120 nm node

120 nm	Emissions to air		
kg/wafer	Expected value	Low	High
VOC			
DMA	1.28E−06	0.00E+00	0.00E+00
MMA	2.95E−05	0.00E+00	0.00E+00
TDMAT	0.00E+00	3.04E−01	1.21E+00
TMS	0.00E+00	0.00E+00	0.00E+00
TEOS	0.00E+00	0.00E+00	0.00E+00
formaldehyde	1.42E−04	1.54E+01	6.16E+01
m-cresol	9.33E−06	9.77E+00	3.91E+01
p-cresol	4.54E−04	9.68E+00	3.87E+01
PGMEA	1.67E−02	1.25E+02	4.99E+02
n-methyl-2-pyrollidone	6.41E−04	5.58E+00	2.23E+01
PGME	1.21E−02	1.64E+02	6.54E+02
ethyl lactate	1.42E−02	1.56E+02	6.25E+02
TMAH	6.19E−04	1.63E+01	6.51E+01
TDEAH	0.00E+00	0.00E+00	0.00E+00
TDMAS	0.00E+00	0.00E+00	0.00E+00
DEA	0.00E+00	0.00E+00	0.00E+00
PDMAT	0.00E+00	0.00E+00	0.00E+00
Flammable			
SiH_4	8.51E−04	6.19E+00	2.48E+01
H_2	3.76E−03	4.50E+01	1.80E+02
DCS	0.00E+00	0.00E+00	0.00E+00
B_2H_6	0.00E+00	0.00E+00	0.00E+00
C_4F_8	1.20E−05	0.00E+00	0.00E+00
GeH_4	0.00E+00	0.00E+00	0.00E+00
Other chemicals			
Ar	6.15E−01	3.11E+02	1.25E+03
N_2	6.78E−01	3.62E+02	1.45E+03
He	4.50E−04	2.25E−01	8.99E−01
O_2	9.23E−02	1.50E+02	6.01E+02
IPA	3.42E−02	3.91E+02	1.57E+03
NH_3	4.50E−04	2.25E−01	8.99E−01
Ti	5.12E−07	2.84E−04	1.14E−03
Cu	7.06E−07	3.92E−04	1.57E−03
Sn	1.68E−05	9.34E−03	3.74E−02

Table B.15 Flash memory manufacturing emissions per wafer, 90 nm node

90 nm	Emissions to air		
kg/wafer	Expected value	Low	High
GWG			
CF_4	1.30E−04	2.94E+00	1.18E+01
CHF_3	2.37E−05	1.65E+00	6.61E+00
C_2F_6	0.00E+00	0.00E+00	0.00E+00
CH_4	2.01E−04	2.66E+02	1.06E+03
CO_2	1.46E+00	0.00E+00	0.00E+00
NF_3	3.62E−04	7.36E+01	2.95E+02
C_4F_6	0.00E+00	0.00E+00	0.00E+00
N_2O	0.00E+00	0.00E+00	0.00E+00
Hazardous			
F_2	3.93E−03	0.00E+00	0.00E+00
HF (gas)	6.04E−04	0.00E+00	0.00E+00
SiF_4	3.67E−05	5.42E−02	2.17E−01
C_4F_8	9.04E−07	4.97E−01	1.99E+00
C_2F_4	1.07E−05	0.00E+00	0.00E+00
COF_2	5.02E−05	0.00E+00	0.00E+00
HCl (gas)	1.31E−07	0.00E+00	0.00E+00
Cl_2	6.07E−05	1.31E+00	5.25E+00
$SiCl_4$	9.01E−07	0.00E+00	0.00E+00
PH_3	0.00E+00	1.33E+00	5.34E+00
bis tertiary-butylamino silane	1.52E−05	9.65E−01	3.86E+00
AsH_3	4.29E−06	1.38E−01	5.50E−01
BCl_3	1.84E−05	6.40E−01	2.56E+00
Pb	0.00E+00	0.00E+00	0.00E+00
WF_6	4.32E−03	1.64E+01	6.54E+01
HBr	0.00E+00	3.99E+00	1.59E+01
Br_2	0.00E+00	0.00E+00	0.00E+00
HMDS	0.00E+00	1.30E+02	5.18E+02
OMCTS	0.00E+00	0.00E+00	0.00E+00
Criteria			
CO	2.18E−02	0.00E+00	0.00E+00
NO	4.62E−03	0.00E+00	0.00E+00
NO_2	4.52E−04	0.00E+00	0.00E+00
SO_2	0.00E+00	0.00E+00	0.00E+00
N_2O	2.65E−04	3.48E+00	1.39E+01
O_3	0.00E+00	0.00E+00	0.00E+00

Table B.16 Flash memory manufacturing emissions per wafer, 90 nm node

90 nm	Emissions to air		
kg/wafer	Expected value	Low	High
VOC			
DMA	1.28E−06	0.00E+00	0.00E+00
MMA	2.95E−05	0.00E+00	0.00E+00
TDMAT	0.00E+00	3.04E−01	1.21E+00
TMS	0.00E+00	0.00E+00	0.00E+00
TEOS	0.00E+00	0.00E+00	0.00E+00
formaldehyde	1.42E−04	1.54E+01	6.16E+01
m-cresol	9.33E−06	9.77E+00	3.91E+01
p-Cresol	4.54E−04	9.68E+00	3.87E+01
PGMEA	1.67E−02	1.25E+02	4.99E+02
n-methyl-2-pyrollidone	6.41E−04	5.58E+00	2.23E+01
PGME	1.21E−02	1.64E+02	6.54E+02
ethyl lactate	1.42E−02	1.56E+02	6.25E+02
TMAH	6.19E−04	1.63E+01	6.51E+01
TDEAH	0.00E+00	0.00E+00	0.00E+00
TDMAS	0.00E+00	0.00E+00	0.00E+00
DEA	0.00E+00	0.00E+00	0.00E+00
PDMAT	0.00E+00	0.00E+00	0.00E+00
Flammable			
SiH_4	8.51E−04	6.19E+00	2.48E+01
H_2	3.76E−03	4.50E+01	1.80E+02
DCS	0.00E+00	0.00E+00	0.00E+00
B_2H_6	0.00E+00	0.00E+00	0.00E+00
C_4F_8	1.20E−05	0.00E+00	0.00E+00
GeH_4	0.00E+00	0.00E+00	0.00E+00
Other chemicals			
Ar	6.15E−01	3.11E+02	1.25E+03
N_2	6.78E−01	3.62E+02	1.45E+03
He	4.50E−04	2.25E−01	8.99E−01
O_2	9.23E−02	1.50E+02	6.01E+02
IPA	3.42E−02	3.91E+02	1.57E+03
NH_3	4.50E−04	2.25E−01	8.99E−01
Ti	5.12E−07	2.84E−04	1.14E−03
Cu	1.41E−06	7.85E−04	3.14E−03
Sn	5.89E−05	2.97E−02	1.19E−01

Table B.17 Flash memory manufacturing emissions per wafer, 65 nm node

65 nm	Emissions to air		
kg/wafer	Expected value	Low	High
GWG			
CF_4	1.63E−04	8.13E−05	3.25E−04
CHF_3	3.66E−05	1.83E−05	7.31E−05
C_2F_6	0.00E+00	0.00E+00	0.00E+00
CH_4	2.11E−04	1.05E−04	4.22E−04
CO_2	1.53E+00	7.66E−01	3.07E+00
NF_3	3.93E−04	1.96E−04	7.85E−04
C_4F_6	0.00E+00	0.00E+00	0.00E+00
N_2O	0.00E+00	0.00E+00	0.00E+00
Hazardous			
F_2	4.51E−03	2.25E−03	9.02E−03
HF (gas)	8.45E−04	4.22E−04	1.69E−03
SiF_4	4.53E−05	2.27E−05	9.06E−05
C_4F_8	1.53E−06	7.67E−07	3.07E−06
C_2F_4	1.81E−05	9.05E−06	3.62E−05
COF_2	4.55E−05	2.28E−05	9.10E−05
HCl (gas)	2.47E−07	1.24E−07	4.95E−07
Cl_2	1.03E−04	5.13E−05	2.05E−04
$SiCl_4$	1.70E−06	8.50E−07	3.40E−06
PH_3	0.00E+00	0.00E+00	0.00E+00
bis tertiary-butylamino silane	1.43E−05	7.17E−06	2.87E−05
AsH_3	4.05E−06	2.02E−06	8.09E−06
BCl_3	3.46E−05	1.73E−05	6.93E−05
Pb	0.00E+00	0.00E+00	0.00E+00
WF_6	8.16E−03	4.08E−03	1.63E−02
HBr	0.00E+00	0.00E+00	0.00E+00
Br_2	0.00E+00	0.00E+00	0.00E+00
HMDS	0.00E+00	0.00E+00	0.00E+00
OMCTS	0.00E+00	0.00E+00	0.00E+00
Criteria			
CO	2.10E−02	1.05E−02	4.21E−02
NO	5.02E−03	2.51E−03	1.00E−02
NO_2	4.58E−04	2.29E−04	9.17E−04
SO_2	0.00E+00	0.00E+00	0.00E+00
N_2O	2.50E−04	1.25E−04	5.00E−04

Table B.18 Flash memory manufacturing emissions per wafer, 65 nm node

65 nm	Emissions to air		
kg/wafer	Expected value	Low	High
VOC			
DMA	2.42E−06	1.21E−06	4.84E−06
MMA	5.57E−05	2.78E−05	1.11E−04
TDMAT	0.00E+00	0.00E+00	0.00E+00
TMS	0.00E+00	0.00E+00	0.00E+00
TEOS	0.00E+00	0.00E+00	0.00E+00
formaldehyde	1.44E−04	7.21E−05	2.88E−04
m-cresol	9.45E−06	4.72E−06	1.89E−05
p-cresol	4.29E−04	2.15E−04	8.58E−04
PGMEA	1.69E−02	8.47E−03	3.39E−02
n-methyl-2-pyrollidone	6.05E−04	3.02E−04	1.21E−03
PGME	1.23E−02	6.14E−03	2.46E−02
ethyl lactate	1.45E−02	7.23E−03	2.89E−02
TMAH	5.95E−04	2.98E−04	1.19E−03
TDEAH	0.00E+00	0.00E+00	0.00E+00
TDMAS	0.00E+00	0.00E+00	0.00E+00
DEA	0.00E+00	0.00E+00	0.00E+00
PDMAT	0.00E+00	0.00E+00	0.00E+00
Flammable			
SiH_4	1.02E−03	5.12E−04	2.05E−03
H_2	8.06E−03	4.03E−03	1.61E−02
DCS	0.00E+00	0.00E+00	0.00E+00
B_2H_6	0.00E+00	0.00E+00	0.00E+00
C_4F_8	1.14E−05	5.68E−06	2.27E−05
GeH_4	0.00E+00	0.00E+00	0.00E+00
Other chemicals			
Ar	7.57E−01	3.79E−01	1.51E+00
N_2	8.27E−01	4.13E−01	1.65E+00
He	8.48E−04	4.24E−04	1.70E−03
O_2	9.94E−02	4.97E−02	1.99E−01
IPA	3.46E−02	1.73E−02	6.91E−02
NH_3	8.48E−04	4.24E−04	1.70E−03
Ti	4.8295E−07	2.41475E−07	9.659E−07
Cu	1.33269E−06	6.66344E−07	2.66537E−06
Sn	0	0	0

Table B.19 Flash memory manufacturing emissions per wafer, 45 nm node

45 nm	Emissions to air		
kg/wafer	Expected value	Low	High
GWG			
CF_4	8.57E−04	1.07E−03	4.28E−03
CHF_3	2.63E−05	6.30E−06	2.52E−05
C_2F_6	3.64E−06	2.50E−06	1.00E−05
CH_4	6.13E−04	2.23E−03	8.94E−03
CO_2	1.51E+00	8.89E−01	3.56E+00
NF_3	4.33E−04	2.15E−04	8.59E−04
C_4F_6	1.07E−05	8.01E−06	3.20E−05
N_2O	0.00E+00	2.50E−04	1.00E−03
Hazardous			
F_2	4.10E−03	1.90E−03	7.61E−03
HF (gas)	1.03E−02	4.62E−03	1.85E−02
SiF_4	8.47E−05	4.43E−05	1.77E−04
C_4F_8	2.78E−06	1.02E−06	4.09E−06
C_2F_4	2.21E−05	4.02E−06	1.61E−05
COF_2	4.72E−05	2.83E−05	1.13E−04
HCl (gas)	1.24E−07	0.00E+00	0.00E+00
Cl_2	6.31E−05	1.17E−05	4.69E−05
$SiCl_4$	8.50E−07	0.00E+00	0.00E+00
PH_3	0.00E+00	1.42E−08	5.68E−08
bis tertiary-butylamino silane	0.00E+00	2.39E−06	9.57E−06
AsH_3	4.05E−06	1.78E−09	7.10E−09
BCl_3	1.73E−05	0.00E+00	0.00E+00
Pb	0.00E+00	0.00E+00	0.00E+00
WF_6	6.12E−03	2.04E−03	8.16E−03
HBr	0.00E+00	0.00E+00	0.00E+00
Br_2	0.00E+00	0.00E+00	0.00E+00
HMDS	0.00E+00	0.00E+00	0.00E+00
OMCTS	0.00E+00	0.00E+00	0.00E+00
Criteria			
CO	2.89E−01	3.14E−01	1.26E+00
NO	7.40E−03	3.49E−03	1.40E−02
NO_2	3.64E−03	3.34E−03	1.34E−02
SO_2	0.00E+00	0.00E+00	0.00E+00
N_2O	4.92E−04	2.50E−04	1.00E−03

Table B.20 Flash memory manufacturing emissions per wafer, 45 nm node

45 nm	Emissions to air		
kg/wafer	Expected value	Low	High
VOC			
DMA	1.21E−06	1.21E−06	4.84E−06
MMA	2.78E−05	2.78E−05	1.11E−04
TDMAT	0.00E+00	0.00E+00	0.00E+00
TMS	0.00E+00	9.26E−04	3.70E−03
TEOS	0.00E+00	6.45E−04	2.58E−03
formaldehyde	2.01E−04	7.97E−05	3.19E−04
m-cresol	1.30E−05	5.05E−06	2.02E−05
p-Cresol	4.33E−04	5.00E−06	2.00E−05
PGMEA	2.36E−02	9.37E−03	3.75E−02
n-methyl-2-pyrollidone	6.05E−04	0.00E+00	0.00E+00
PGME	1.69E−02	6.77E−03	2.71E−02
ethyl lactate	2.01E−02	8.00E−03	3.20E−02
TMAH	6.58E−04	8.81E−05	3.52E−04
TDEAH	0.00E+00	0.00E+00	0.00E+00
TDMAS	0.00E+00	0.00E+00	0.00E+00
DEA	0.00E+00	0.00E+00	0.00E+00
PDMAT	0.00E+00	0.00E+00	0.00E+00
Flammable			
SiH_4	9.64E−04	2.20E−04	8.80E−04
H_2	6.28E−03	2.53E−03	1.01E−02
DCS	0.00E+00	0.00E+00	0.00E+00
B_2H_6	0.00E+00	0.00E+00	0.00E+00
C_4F_8	1.70E−04	1.65E−04	6.59E−04
GeH_4	0.00E+00	0.00E+00	0.00E+00
Other chemicals			
Ar	1.69E+00	9.55E−01	3.82E+00
N_2	6.31E−01	7.60E−01	3.04E+00
He	4.24E−04	7.44E−02	2.98E−01
O_2	1.01E−01	2.93E−02	1.17E−01
IPA	4.72E−02	1.83E−02	7.30E−02
NH_3	4.24E−04	7.44E−02	2.98E−01
Ti	4.83E−07	4.83E−08	1.93E−07
Cu	1.33E−06	2.61E−05	1.04E−04
Sn	5.55E−05	2.17E−03	8.67E−03

B.3 Flash LCA: Tabulated Results with Uncertainty

Table B.21 Flash life cycle impacts per wafer, 45 nm node

	Photochemical smog kg NO_x			Acidification mol H^+			Ecotoxicity kg 2,4-D		
	Expected value	Lower bound	Upper bound	Expected value	Lower bound	Upper bound	Expected value	Lower bound	Upper bound
Transportation	0.22	N/A	N/A	227	N/A	N/A	0.00	N/A	N/A
Electricity (Si, fab. and use)	0.21	0.20	0.23	126	117	134	1.06	1.03	1.08
Fab. direct emissions	0.089	0.045	0.18	65	32	129	5.8E−02	5.1E−02	6.5E−02
Infrastructure	0.99	0.49	1.24	1,133	566	1,416	7.6E−03	3.8E−03	9.5E−03
Total	1.51	0.96	1.86	1,549	942	1,906	1.12	1.09	1.2

Table B.22 Flash life cycle impacts per wafer, 45 nm node

	Human health cancer kg C_6H_6			Human health criteria total DALYs			Health noncancer kg C_7H_7		
	Expected value	Lower bound	Upper bound	Expected value	Lower bound	Upper bound	Expected value	Lower bound	Upper bound
Transportation	0.00	N/A	N/A	0.00	N/A	N/A	0.05	N/A	N/A
Electricity (Si, fab. and use)	0.00E+00	0.00E+00	0.00E+00	3.48E-03	3.41E-03	3.55E-03	85	83	87
Fab. direct emissions	2.25E-02	1.47E-02	3.70E-02	3.16E-05	1.58E-05	6.32E-05	864	760	970
Infrastructure	1.12E-02	5.61E-03	1.40E-02	1.73E-02	8.67E-03	2.17E-02	476	238	595
Total	0.034	0.020	0.051	0.022	0.013	0.026	1,425	1,080	1,652

	Eutrophication, to air kg N			Eutrophication, to water kg N		
	Expected value	Lower bound	Upper bound	Expected value	Lower bound	Upper bound
Transportation	7.69E-03	N/A	N/A	N/A	N/A	N/A
Electricity (Si, fab. and use)	7.67E-03	7.29E-03	8.06E-03	4.63	2.36	9.15
Fab. direct emissions	6.36E-04	3.18E-04	1.27E-03	N/A	N/A	N/A
Infrastructure	0.034	0.017	0.042	N/A	N/A	N/A
Total	0.050	0.032	0.059	4.6	2.4	9.2

Table B.23 Flash life cycle impacts per wafer, 65 nm node

	Photochemical smog kg NO_x			Acidification mol H^+			Ecotoxicity kg 2,4-D		
	Expected value	Lower bound	Upper bound	Expected value	Lower bound	Upper bound	Expected value	Lower bound	Upper bound
Transportation	0.22	N/A	N/A	234	N/A	N/A	0.00	N/A	N/A
Electricity (Si, fab. and use)	0.19	0.18	0.20	100	91	108	0.92	0.89	0.94
Fab. direct emissions	0.054	0.027	0.11	17	8	34	2.1E−03	1.5E−03	3.3E−03
Infrastructure	0.78	0.39	0.98	894	447	1,117	5.8E−03	2.9E−03	7.3E−03
Total	1.24	0.81	1.50	1,244	781	1,493	0.93	0.90	1.0

Table B.24 Flash life cycle impacts per wafer, 65 nm node

	Human health cancer kg C_6H_6			Human health criteria total DALYs			Health noncancer kg C_7H_7		
	Expected value	Lower bound	Upper bound	Expected value	Lower bound	Upper bound	Expected value	Lower bound	Upper bound
Transportation	0.00	N/A	N/A	0.00	N/A	N/A	0.06	N/A	N/A
Electricity (Si, fab. and use)	0.00E+00	0.00E+00	0.00E+00	3.50E−03	3.43E−03	3.57E−03	74	72	75
Fab. direct emissions	2.22E−02	1.44E−02	3.66E−02	1.57E−05	7.84E−06	3.13E−05	777	683	873
Infrastructure	8.62E−03	4.31E−03	1.08E−02	1.37E−02	6.86E−03	1.71E−02	365	183	457
Total	0.031	0.019	0.047	0.019	0.012	0.022	1,216	937	1,406

	Eutrophication, to air kg N			Eutrophication, to water kg N		
	Expected value	Lower bound	Upper bound	Expected value	Lower bound	Upper bound
Transportation	7.69E−03	N/A	N/A	N/A	N/A	N/A
Electricity (Si, fab. and use)	6.66E−03	6.31E−03	7.02E−03	N/A	N/A	N/A
Fab. direct emissions	3.66E−04	1.83E−04	7.33E−04	4.47	2.28	8.84
Infrastructure	0.027	0.013	0.033	N/A	N/A	N/A
Total	0.041	0.027	0.049	4.5	2.3	8.8

Table B.25 Flash life cycle impacts per wafer, 90 nm node

	Photochemical smog kg NO_x			Acidification mol H^+			Ecotoxicity kg 2,4-D		
	Expected value	Lower bound	Upper bound	expected value	Lower bound	Upper bound	expected value	Lower bound	Upper bound
Transportation	0.20	N/A	N/A	211	N/A	N/A	0.00	N/A	N/A
Electricity (Si, fab. and use)	0.18	0.17	0.19	100	92	107	0.89	0.87	0.91
Fab. direct emissions	0.05	0.026	0.10	15	7.3	29	2.3E−03	1.4E−03	3.8E−03
Infrastructure	0.61	0.31	0.77	703	352	879	4.5E−03	2.2E−03	5.6E−03
Total	1.05	0.71	1.27	1,029	662	1,227	0.90	0.87	0.9

Table B.26 Flash life cycle impacts per wafer, 90 nm node

	Human health cancer kg C_6H_6			Human health criteria total DALYs			Health noncancer kg C_7H_7		
	Expected value	Lower bound	Upper bound	Expected value	Lower bound	Upper bound	Expected value	Lower bound	Upper bound
Transportation	0.00	N/A	N/A	0.00	N/A	N/A	0.05	N/A	N/A
Electricity (Si, fab. and use)	0.00E+00	0.00E+00	0.00E+00	3.08E−03	3.02E−03	3.15E−03	72	70	73
Fab. direct emissions	2.23E−02	1.42E−02	3.75E−02	1.45E−05	7.26E−06	2.90E−05	751	660	843
Infrastructure	6.61E−03	3.31E−03	8.27E−03	1.08E−02	5.41E−03	1.35E−02	281	140	351
Total	0.029	0.018	0.046	0.015	0.010	0.018	1,103	870	1,267

	Eutrophication, to air kg N			Eutrophication, to water kg N		
	Expected value	Lower bound	Upper bound	Expected value	Lower bound	Upper bound
Transportation	7.17E−03	N/A	N/A	N/A	N/A	N/A
Electricity (Si, fab. and use)	6.48E−03	6.14E−03	6.82E−03	N/A	N/A	N/A
Fab. direct emissions	3.34E−04	1.67E−04	6.67E−04	4.13	2.10	8.16
Infrastructure	0.021	0.010	0.026	N/A	N/A	N/A
Total	0.035	0.024	0.041	4.1	2.1	8.2

Table B.27 Flash life cycle impacts per wafer, 120 nm node

	Photochemical smog kg NO_x			Acidification mol H^+			Ecotoxicity kg 2,4-D		
	Expected value	Lower bound	Upper bound	Expected value	Lower bound	Upper bound	Expected value	Lower bound	Upper bound
Transportation	0.25	N/A	N/A	264	N/A	N/A	0.00	N/A	N/A
Electricity (Si, fab. and use)	0.18	0.17	0.19	78	70	86	0.86	0.84	0.88
Fab. direct emissions	0.052	0.026	0.10	14	7.1	29	3.3E−03	2.0E−03	5.9E−03
Infrastructure	0.56	0.28	0.70	638	319	798	4.1E−03	2.0E−03	5.1E−03
Total	1.04	0.72	1.24	994	660	1,177	0.87	0.85	0.90

Table B.28 Flash life cycle impacts per wafer, 120 nm node

	Human health cancer kg C_6H_6			Human health criteria total DALYs			Health noncancer kg C_7H_7		
	Expected value	Lower bound	Upper bound	Expected value	Lower bound	Upper bound	Expected value	Lower bound	Upper bound
Transportation	0.00	N/A	N/A	0.00	N/A	N/A	0.06	N/A	N/A
Electricity (Si, fab. and use)	0.00E+00	0.00E+00	0.00E+00	3.40E−03	3.33E−03	3.47E−03	69	68	71
Fab. direct emissions	1.96E−02	1.10E−02	3.64E−02	1.40E−05	7.00E−06	2.80E−05	842	706	1,026
Infrastructure	6.06E−03	3.03E−03	7.57E−03	9.81E−03	4.91E−03	1.23E−02	257	128	321
Total	0.026	0.014	0.044	0.015	0.010	0.017	1,169	902	1,418

	Eutrophication, to air kg N			Eutrophication, to water kg N		
	Expected value	Lower bound	Upper bound	Expected value	Lower bound	Upper bound
Transportation	8.95E−03	N/A	N/A	N/A	N/A	N/A
Electricity (Si, fab. and use)	6.27E−03	5.90E−03	6.64E−03	N/A	N/A	N/A
Fab. direct emissions	3.23E−04	1.61E−04	6.45E−04	4.19	2.11	8.34
Infrastructure	0.019	0.009	0.024	N/A	N/A	N/A
Total	0.034	0.024	0.040	4.2	2.1	8.3

Table B.29 Flash life cycle impacts per wafer, 150 nm node

	Photochemical smog kg NO$_x$			Acidification mol H$^+$			Ecotoxicity kg 2,4-D		
	Expected value	Lower bound	Upper bound	Expected value	Lower bound	Upper bound	Expected value	Lower bound	Upper bound
Transportation	0.25	N/A	N/A	264	N/A	N/A	0.00	N/A	N/A
Electricity (Si, fab. and use)	0.20	0.19	0.21	98	89	106	0.97	0.94	0.99
Fab. direct emissions	0.072	0.036	0.14	33	17	67	3.4E–03	2.0E–03	6.1E–03
Infrastructure	0.39	0.20	0.49	450	225	562	2.9E–03	1.4E–03	3.6E–03
Total	0.92	0.67	1.10	845	594	999	0.97	0.95	1.0

Table B.30 Flash life cycle impacts per wafer, 150 nm node

	Human health cancer kg C_6H_6			Human health criteria total DALYs			Health noncancer kg C_7H_7		
	Expected value	Lower bound	Upper bound	Expected value	Lower bound	Upper bound	Expected value	Lower bound	Upper bound
Transportation	0.00	N/A	N/A	0.00	N/A	N/A	0.06	N/A	N/A
Electricity (Si, fab. and use)	0.00E+00	0.00E+00	0.00E+00	3.59E−03	3.52E−03	3.66E−03	78	76	80
Fab. direct emissions	1.93E−02	1.10E−02	3.55E−02	3.09E−05	1.54E−05	6.17E−05	896	751	1,092
Infrastructure	4.23E−03	2.12E−03	5.29E−03	6.92E−03	3.46E−03	8.65E−03	179	90	224
Total	0.024	0.013	0.041	0.012	0.008	0.014	1,154	917	1,396

	Eutrophication, to air kg N			Eutrophication, to water kg N		
	Expected value	Lower bound	Upper bound	Expected value	Lower bound	Upper bound
Transportation	8.95E−03	N/A	N/A	N/A	N/A	N/A
Electricity (Si, fab. and use)	7.04E−03	6.65E−03	7.42E−03	N/A	N/A	N/A
Fab. direct emissions	6.17E−04	3.08E−04	1.23E−03	4.46	2.25	8.88
Infrastructure	0.013	0.007	0.017	N/A	N/A	N/A
Total	0.030	0.023	0.034	4.5	2.2	8.9

Appendix C
DRAM Life Cycle Inventory and Impact Results

C.1 DRAM Manufacturing Inventory Data

Table C.1 DRAM manufacturing inventory per wafer, 250 nm node

250 nm		250 nm	
Inputs	g/wafer	Inputs	g/wafer
GWG		*Flammable*	
CF_4	9.37E−03	SiH_4	1.23E−02
CHF_3	9.60E−04	H_2	1.72E−01
C_2F_6	0.00E+00	DCS	0.00E+00
CH_4	2.36E−01	B_2H_6	0.00E+00
CO_2	0.00E+00	C_4F_8	0.00E+00
NF_3	1.75E−01	GeH_4	0.00E+00
C_4F_6	0.00E+00	C_2H_2	0.00E+00
N_2O	0.00E+00	C_2H_4	0.00E+00
Hazardous			
F_2	0.00E+00	*Other chemicals*	
HF (gas)	0.00E+00	O_2	1.81E+00
SiF_4	6.14E−03	Ar	5.35E+00
C_4F_8	1.36E−03	N_2	1.49E+00
C_2F_4	0.00E+00	CMP polishing solution	0.00E+00
COF_2	0.00E+00	H_2O_2	1.09E+01
HCl (gas)	0.00E+00	abrasive	0.00E+00
Cl_2	3.63E−03	silica slurry	0.00E+00
$SiCl_4$	0.00E+00	benzotriazole	0.00E+00
PH_3	9.13E−04	CMP surfactants	1.75E−01

(continued)

S.B. Boyd, *Life-Cycle Assessment of Semiconductors*, DOI 10.1007/978-1-4419-9988-7,   179
© Springer Science+Business Media, LLC 2012

Table C.1 (continued)

250 nm		250 nm	
Inputs	g/wafer	Inputs	g/wafer
BTBAS	0.00E+00	citric acid	9.74E−02
AsH$_3$	0.00E+00	tungsten CMP slurry	8.43E+00
BCl$_3$	3.88E−03	Al	2.61E−04
Pb	1.63E−03	As	0.00E+00
WF$_6$	6.20E−02	Cu	2.56E−04
HBr	4.03E−03	Pt	0.00E+00
Br$_2$	0.00E+00	Ti	8.62E−08
HMDS	1.61E−01	He	3.07E−03
OMCTS	0.00E+00	NH$_3$	3.57E−04

Table C.2 DRAM manufacturing inventory per wafer, 250 nm node

250 nm		250 nm	
Inputs	g/wafer	Inputs	g/wafer
Criteria		*Other chemicals*	
CO	0.00E+00	CuSO$_4$	0.00E+00
NO	0.00E+00	ArH	0.00E+00
NO$_2$	0.00E+00	100% HF liquid	6.62E−02
SO$_2$	0.00E+00	H$_2$SO$_4$	1.26E+02
N$_2$O	5.27E−03	HCl	2.15E+00
O$_3$	0.00E+00	NH$_4$OH	3.76E+00
		IPA	4.15E+00
VOC		oxide CMP slurry, chemicals	1.62E+01
DMA	0.00E+00	BF$_3$	0.00E+00
MMA	0.00E+00	Cr	0.00E+00
TDMAT	4.14E−03	Au	7.38E−05
TMS	0.00E+00	Sn	4.52E−04
TEOS	0.00E+00	Co	3.94E−05
formaldehyde	1.92E−02	Ni	3.31E−03
m-cresol	1.22E−02	polyamide laminate	1.89E−03
p-cresol	1.21E−02	laminate solvent	0.00E+00
PGMEA	1.55E−01	H$_3$PO$_4$	0.00E+00
n-methyl-2-pyrollidone	8.47E−03	CuCl$_2$	0.00E+00
PGME	2.04E−01	W	3.86E−07
ethyl lactate	1.95E−01	Ta	0.00E+00
TMAH	2.10E−02	Cu ECP additive X	0.00E+00
TDEAH	0.00E+00	Cu ECP Additive Y	0.00E+00
TDMAS	0.00E+00	Na$_2$B$_4$O$_7$	9.08E−03
DEA	0.00E+00		
PDMAT	0.00E+00		

Table C.3 DRAM manufacturing inventory per wafer, 180 nm node

180 nm		180 nm	
Inputs	g/wafer	Inputs	g/wafer
GWG		*Flammable*	
CF_4	9.37E−03	SiH_4	1.23E−02
CHF_3	9.60E−04	H_2	1.72E−01
C_2F_6	0.00E+00	DCS	0.00E+00
CH_4	2.36E−01	B_2H_6	0.00E+00
CO_2	0.00E+00	C_4F_8	0.00E+00
NF_3	1.75E−01	GeH_4	0.00E+00
C_4F_6	0.00E+00	C_2H_2	0.00E+00
N_2O	0.00E+00	C_2H_4	0.00E+00
Hazardous			
F_2	0.00E+00	*Other chemicals*	
HF (gas)	0.00E+00	O_2	1.81E+00
SiF_4	6.14E−03	Ar	5.35E+00
C_4F_8	1.36E−03	N_2	1.49E+00
C_2F_4	0.00E+00	CMP polishing solution	0.00E+00
COF_2	0.00E+00	H_2O_2	1.09E+01
HCl (gas)	0.00E+00	abrasive	0.00E+00
Cl_2	3.63E−03	silica slurry	0.00E+00
$SiCl_4$	0.00E+00	benzotriazole	0.00E+00
PH_3	9.13E−04	CMP surfactants	1.75E−01
BTBAS	0.00E+00	citric acid	9.74E−02
AsH_3	0.00E+00	tungsten CMP slurry	8.43E+00
BCl_3	3.88E−03	Al	2.61E−04
Pb	1.20E−03	As	0.00E+00
WF_6	6.20E−02	Cu	1.87E−04
HBr	4.03E−03	Pt	0.00E+00
Br_2	0.00E+00	Ti	8.62E−08
HMDS	1.61E−01	He	3.07E−03
OMCTS	0.00E+00	NH_3	3.57E−04

Table C.4 DRAM manufacturing inventory per wafer, 180 nm node

180 nm		180 nm	
Inputs	g/wafer	Inputs	g/wafer
Criteria		*Other chemicals*	
CO	0.00E+00	$CuSO_4$	0.00E+00
NO	0.00E+00	ArH	0.00E+00
NO_2	0.00E+00	100% HF liquid	6.62E−02
SO_2	0.00E+00	H_2SO_4	1.26E+02
N_2O	5.27E−03	HCl	2.15E+00
O_3	0.00E+00	NH_4OH	3.76E+00
		IPA	4.15E+00

(continued)

Table C.4 (continued)

180 nm		180 nm	
Inputs	g/wafer	Inputs	g/wafer
VOC		oxide CMP slurry, chemicals	1.62E+01
DMA	0.00E+00	BF_3	0.00E+00
MMA	0.00E+00	Cr	0.00E+00
TDMAT	4.14E−03	Au	5.39E−05
TMS	0.00E+00	Sn	3.31E−04
TEOS	0.00E+00	Co	2.88E−05
formaldehyde	1.92E−02	Ni	2.42E−03
m-cresol	1.22E−02	polyamide laminate	1.89E−03
p-Cresol	1.21E−02	laminate solvent	0.00E+00
PGMEA	1.55E−01	H_3PO_4	0.00E+00
n-methyl-2-pyrollidone	8.47E−03	$CuCl_2$	0.00E+00
PGME	2.04E−01	W	3.86E−07
ethyl lactate	1.95E−01	Ta	0.00E+00
TMAH	2.10E−02	Cu ECP additive X	0.00E+00
TDEAH	0.00E+00	Cu ECP Additive Y	0.00E+00
TDMAS	0.00E+00	$Na_2B_4O_7$	9.08E−03
DEA	0.00E+00		
PDMAT	0.00E+00		

Table C.5 DRAM manufacturing inventory per wafer, 130 nm node

130 nm		130 nm	
Inputs	g/wafer	Inputs	g/wafer
GWG		*Flammable*	
CF_4	2.07E−03	SiH_4	1.63E−02
CHF_3	1.03E−03	H_2	2.27E−01
C_2F_6	0.00E+00	DCS	0.00E+00
CH_4	2.59E−01	B_2H_6	0.00E+00
CO_2	0.00E+00	C_4F_8	3.75E−04
NF_3	2.31E−01	GeH_4	0.00E+00
C_4F_6	7.23E−05	C_2H_2	0.00E+00
N_2O	0.00E+00	C_2H_4	0.00E+00
Hazardous			
F_2	0.00E+00	*Other chemicals*	
HF (gas)	0.00E+00	O_2	2.28E+00
SiF_4	8.10E−03	Ar	7.05E+00
C_4F_8	1.79E−03	N_2	1.96E+00
C_2F_4	0.00E+00	CMP polishing solution	0.00E+00
COF_2	0.00E+00	H_2O_2	1.47E+01
HCl (gas)	0.00E+00	abrasive	0.00E+00
Cl_2	1.14E−03	silica slurry	0.00E+00
$SiCl_4$	0.00E+00	benzotriazole	0.00E+00
PH_3	1.20E−03	CMP surfactants	2.31E−01

(continued)

Table C.5 (continued)

130 nm		130 nm	
Inputs	g/wafer	Inputs	g/wafer
BTBAS	0.00E+00	citric acid	1.29E−01
AsH$_3$	0.00E+00	tungsten CMP slurry	1.11E+01
BCl$_3$	0.00E+00	Al	0.00E+00
Pb	5.07E−03	As	0.00E+00
WF$_6$	8.18E−02	Cu	7.97E−04
HBr	5.32E−03	Pt	0.00E+00
Br$_2$	0.00E+00	Ti	1.14E−07
HMDS	2.31E−01	He	4.05E−03
OMCTS	0.00E+00	NH$_3$	4.71E−04

Table C.6 DRAM manufacturing inventory per wafer, 130 nm node

130 nm		130 nm	
Inputs	g/wafer	Inputs	g/wafer
Criteria		*Other chemicals*	
CO	0.00E+00	CuSO$_4$	0.00E+00
NO	0.00E+00	ArH	0.00E+00
NO$_2$	0.00E+00	100% HF liquid	1.03E−01
SO$_2$	0.00E+00	H$_2$SO$_4$	1.67E+02
N$_2$O	6.96E−03	HCl	3.03E+00
O$_3$	0.00E+00	NH$_4$OH	5.19E+00
		IPA	6.26E+00
VOC		oxide CMP slurry, chemicals	2.13E+01
DMA	0.00E+00	BF$_3$	0.00E+00
MMA	0.00E+00	Cr	0.00E+00
TDMAT	5.47E−03	Au	2.30E−04
TMS	0.00E+00	Sn	1.40E−03
TEOS	0.00E+00	Co	1.23E−04
formaldehyde	2.75E−02	Ni	1.03E−02
m-cresol	1.75E−02	polyamide laminate	2.49E−03
p-cresol	1.73E−02	laminate solvent	0.00E+00
PGMEA	2.23E−01	H$_3$PO$_4$	6.02E−03
n-methyl-2-pyrollidone	1.12E−02	CuCl$_2$	0.00E+00
PGME	2.92E−01	W	1.71E−06
ethyl lactate	2.79E−01	Ta	3.25E−01
TMAH	3.01E−02	Cu ECP additive X	0.00E+00
TDEAH	0.00E+00	Cu ECP Additive Y	0.00E+00
TDMAS	0.00E+00	Na$_2$B$_4$O$_7$	1.30E−02
DEA	0.00E+00		
PDMAT	0.00E+00		

Table C.7 DRAM manufacturing inventory per wafer, 90 nm node

90 nm		90 nm	
Inputs	g/wafer	Inputs	g/wafer
GWG		*Flammable*	
CF_4	6.38E−03	SiH_4	2.10E−02
CHF_3	3.50E−03	H_2	1.35E−01
C_2F_6	0.00E+00	DCS	0.00E+00
CH_4	4.89E−01	B_2H_6	0.00E+00
CO_2	0.00E+00	C_4F_8	0.00E+00
NF_3	2.09E−01	GeH_4	0.00E+00
C_4F_6	0.00E+00	C_2H_2	0.00E+00
N_2O	0.00E+00	C_2H_4	0.00E+00
Hazardous			
F_2	0.00E+00	*Other chemicals*	
HF (gas)	0.00E+00	O_2	1.98E+00
SiF_4	5.80E−03	Ar	5.61E+00
C_4F_8	1.79E−03	N_2	3.34E−01
C_2F_4	0.00E+00	CMP polishing solution	0.00E+00
COF_2	0.00E+00	H_2O_2	1.33E+01
HCl (gas)	0.00E+00	abrasive	0.00E+00
Cl_2	2.28E−03	silica slurry	0.00E+00
$SiCl_4$	0.00E+00	benzotriazole	0.00E+00
PH_3	1.81E−03	CMP surfactants	1.73E−01
BTBAS	0.00E+00	citric acid	9.64E−02
AsH_3	0.00E+00	tungsten CMP slurry	8.34E+00
BCl_3	0.00E+00	Al	0.00E+00
Pb	5.07E−03	As	0.00E+00
WF_6	4.91E−02	Cu	7.97E−04
HBr	1.06E−02	Pt	0.00E+00
Br_2	0.00E+00	Ti	1.14E−07
HMDS	2.87E−01	He	4.50E−04
OMCTS	0.00E+00	NH_3	9.79E−04

Table C.8 DRAM manufacturing inventory per wafer, 90 nm node

90 nm		90 nm	
Inputs	g/wafer	Inputs	g/wafer
Criteria		*Other chemicals*	
CO	0.00E+00	$CuSO_4$	0.00E+00
NO	0.00E+00	ArH	0.00E+00
NO_2	0.00E+00	100% HF liquid	1.06E−01
SO_2	0.00E+00	H_2SO_4	1.15E+02
N_2O	1.39E−02	HCl	4.16E+00
O_3	0.00E+00	NH_4OH	6.37E+00
		IPA	0.00E+00

(continued)

Table C.8   (continued)

90 nm		90 nm	
Inputs	g/wafer	Inputs	g/wafer
VOC		oxide CMP slurry, chemicals	2.13E+01
DMA	0.00E+00	BF_3	0.00E+00
MMA	0.00E+00	Cr	0.00E+00
TDMAT	6.07E−04	Au	2.30E−04
TMS	2.20E−03	Sn	1.40E−03
TEOS	1.40E−03	Co	1.23E−04
formaldehyde	3.33E−02	Ni	1.03E−02
m-cresol	3.81E−02	polyamide laminate	2.49E−03
p-Cresol	2.00E−02	laminate solvent	0.00E+00
PGMEA	2.82E−01	H_3PO_4	0.00E+00
n-methyl-2-pyrollidone	3.35E−02	$CuCl_2$	1.04E−03
PGME	3.41E−01	W	5.10E−07
ethyl lactate	3.24E−01	Ta	0.00E+00
TMAH	3.49E−02	Cu ECP additive X	3.00E−04
TDEAH	5.67E−05	Cu ECP Additive Y	0.00E+00
TDMAS	0.00E+00	$Na_2B_4O_7$	2.56E−01
DEA	0.00E+00		
PDMAT	0.00E+00		

Table C.9   DRAM manufacturing inventory per wafer, 70 nm node

70 nm		70 nm	
Inputs	g/wafer	Inputs	g/wafer
GWG		*Flammable*	
CF_4	6.16E−03	SiH_4	1.77E−02
CHF_3	3.56E−03	H_2	1.27E−01
C_2F_6	0.00E+00	DCS	0.00E+00
CH_4	5.27E−01	B_2H_6	0.00E+00
CO_2	0.00E+00	C_4F_8	0.00E+00
NF_3	1.61E−01	GeH_4	0.00E+00
C_4F_6	0.00E+00	C_2H_2	0.00E+00
N_2O	0.00E+00	C_2H_4	0.00E+00
Hazardous			
F_2	0.00E+00	*Other chemicals*	
HF (gas)	0.00E+00	O_2	6.52E−01
SiF_4	3.07E−04	Ar	1.49E+00
C_4F_8	2.06E−03	N_2	3.27E−01
C_2F_4	0.00E+00	CMP polishing solution	0.00E+00
COF_2	0.00E+00	H_2O_2	1.31E+01
HCl (gas)	0.00E+00	abrasive	0.00E+00
Cl_2	2.15E−03	silica slurry	0.00E+00
$SiCl_4$	0.00E+00	benzotriazole	0.00E+00
PH_3	4.83E−03	CMP surfactants	1.64E−01

(continued)

Table C.9   (continued)

70 nm		70 nm	
Inputs	g/wafer	Inputs	g/wafer
BTBAS	0.00E+00	citric acid	9.09E−02
AsH$_3$	0.00E+00	tungsten CMP slurry	7.87E+00
BCl$_3$	0.00E+00	Al	0.00E+00
Pb	0.00E+00	As	0.00E+00
WF$_6$	4.63E−02	Cu	6.74E−04
HBr	1.00E−02	Pt	0.00E+00
Br$_2$	0.00E+00	Ti	1.07E−07
HMDS	2.62E−01	He	4.24E−04
OMCTS	0.00E+00	NH$_3$	9.57E−04

Table C.10   DRAM manufacturing inventory per wafer, 70 nm node

70 nm		70 nm	
Inputs	g/wafer	Inputs	g/wafer
Criteria		*Other chemicals*	
CO	0.00E+00	CuSO$_4$	0.00E+00
NO	0.00E+00	ArH	0.00E+00
NO$_2$	0.00E+00	100% HF liquid	1.36E−01
SO$_2$	0.00E+00	H$_2$SO$_4$	1.08E+02
N$_2$O	1.31E−02	HCl	4.28E+00
O$_3$	0.00E+00	NH$_4$OH	6.39E+00
		IPA	0.00E+00
VOC		oxide CMP slurry, chemicals	1.76E+01
DMA	0.00E+00	BF$_3$	0.00E+00
MMA	0.00E+00	Cr	0.00E+00
TDMAT	5.73E−04	Au	1.94E−04
TMS	2.08E−03	Sn	3.77E−03
TEOS	1.32E−03	Co	1.04E−04
formaldehyde	3.04E−02	Ni	8.70E−03
m-cresol	3.52E−02	polyamide laminate	2.35E−03
p-Cresol	1.83E−02	laminate solvent	0.00E+00
PGMEA	2.57E−01	H$_3$PO$_4$	0.00E+00
n-methyl-2-pyrollidone	3.16E−02	CuCl$_2$	9.83E−04
PGME	3.11E−01	W	4.81E−07
ethyl lactate	2.95E−01	Ta	0.00E+00
TMAH	3.18E−02	Cu ECP additive X	2.83E−04
TDEAH	5.35E−05	Cu ECP Additive Y	0.00E+00
TDMAS	0.00E+00	Na$_2$B$_4$O$_7$	2.41E−01
DEA	0.00E+00		
PDMAT	0.00E+00		

Table C.11 DRAM manufacturing inventory per wafer, 57 nm node

57 nm		57 nm	
Inputs	g/wafer	Inputs	g/wafer
GWG		*Flammable*	
CF_4	2.19E−02	SiH_4	1.85E−02
CHF_3	3.41E−03	H_2	2.13E−01
C_2F_6	0.00E+00	DCS	0.00E+00
CH_4	6.48E−01	B_2H_6	0.00E+00
CO_2	0.00E+00	C_4F_8	0.00E+00
NF_3	2.05E−01	GeH_4	0.00E+00
C_4F_6	0.00E+00	C_2H_2	0.00E+00
N_2O	0.00E+00	C_2H_4	0.00E+00
Hazardous			
F_2	0.00E+00	*Other chemicals*	
HF (gas)	0.00E+00	O_2	7.04E−01
SiF_4	2.56E−04	Ar	1.46E+00
C_4F_8	1.31E−03	N_2	5.37E−01
C_2F_4	0.00E+00	CMP polishing solution	0.00E+00
COF_2	0.00E+00	H_2O_2	1.31E+01
HCl (gas)	0.00E+00	abrasive	0.00E+00
Cl_2	3.77E−03	silica slurry	0.00E+00
$SiCl_4$	0.00E+00	benzotriazole	0.00E+00
PH_3	2.49E−03	CMP surfactants	2.18E−01
BTBAS	0.00E+00	citric acid	1.21E−01
AsH_3	6.21E−04	tungsten CMP slurry	1.05E+01
BCl_3	0.00E+00	Al	0.00E+00
Pb	0.00E+00	As	0.00E+00
WF_6	7.71E−02	Cu	1.00E−03
HBr	1.75E−02	Pt	0.00E+00
Br_2	0.00E+00	Ti	1.07E−07
HMDS	2.88E−01	He	8.48E−04
OMCTS	0.00E+00	NH_3	9.23E−04

Table C.12 DRAM manufacturing inventory per wafer, 57 nm node

57 nm		57 nm	
Inputs	g/wafer	Inputs	g/wafer
Criteria		*Other chemicals*	
CO	0.00E+00	$CuSO_4$	0.00E+00
NO	0.00E+00	ArH	0.00E+00
NO_2	0.00E+00	100% HF liquid	1.30E−01
SO_2	0.00E+00	H_2SO_4	9.92E+01
N_2O	1.31E−02	HCl	4.64E+00
O_3	0.00E+00	NH_4OH	6.99E+00
		IPA	0.00E+00

(continued)

Table C.12 (continued)

57 nm		57 nm	
Inputs	g/wafer	Inputs	g/wafer
VOC		oxide CMP slurry, chemicals	1.51E+01
DMA	0.00E+00	BF_3	0.00E+00
MMA	0.00E+00	Cr	0.00E+00
TDMAT	1.15E−03	Au	2.89E−04
TMS	2.08E−03	Sn	5.60E−03
TEOS	1.32E−03	Co	1.54E−04
formaldehyde	3.35E−02	Ni	1.29E−02
m-cresol	3.72E−02	polyamide laminate	2.35E−03
p-Cresol	2.02E−02	laminate solvent	0.00E+00
PGMEA	2.82E−01	H_3PO_4	0.00E+00
n-methyl-2-pyrollidone	3.16E−02	$CuCl_2$	9.83E−04
PGME	3.44E−01	W	4.81E−07
ethyl lactate	3.27E−01	Ta	0.00E+00
TMAH	3.52E−02	Cu ECP additive X	2.83E−04
TDEAH	1.07E−04	Cu ECP Additive Y	0.00E+00
TDMAS	0.00E+00	$Na_2B_4O_7$	2.43E−01
DEA	0.00E+00		
PDMAT	0.00E+00		

C.2   DRAM Manufacturing Emissions Data

Table C.13 DRAM manufacturing emissions per wafer, 250 nm node

250 nm	Emissions to air		
kg/wafer	Expected value	Low	High
GWG			
CF_4	5.20E−04	4.69E−03	1.87E−02
CHF_3	4.60E−05	4.80E−04	1.92E−03
C_2F_6	0.00E+00	0.00E+00	0.00E+00
CH_4	1.46E−04	1.18E−01	4.73E−01
CO_2	6.51E−01	0.00E+00	0.00E+00
NF_3	3.77E−04	8.74E−02	3.50E−01
C_4F_6	0.00E+00	0.00E+00	0.00E+00
N_2O	0.00E+00	0.00E+00	0.00E+00
Hazardous			
F_2	2.98E−03	0.00E+00	0.00E+00
HF (gas)	2.52E−03	0.00E+00	0.00E+00
SiF_4	4.03E−05	3.07E−03	1.23E−02
C_4F_8	1.23E−06	6.78E−04	2.71E−03
C_2F_4	1.45E−05	0.00E+00	0.00E+00
COF_2	4.12E−06	0.00E+00	0.00E+00

(continued)

Table C.13 (continued)

250 nm	Emissions to air		
kg/wafer	Expected value	Low	High
HCl (gas)	3.97E−07	0.00E+00	0.00E+00
Cl_2	1.37E−04	1.82E−03	7.26E−03
$SiCl_4$	2.73E−06	0.00E+00	0.00E+00
PH_3	0.00E+00	4.57E−04	1.83E−03
BTBAS	0.00E+00	0.00E+00	0.00E+00
AsH_3	0.00E+00	0.00E+00	0.00E+00
BCl_3	5.57E−05	1.94E−03	7.77E−03
Pb	1.63E−04	8.17E−04	3.27E−03
WF_6	8.20E−03	3.10E−02	1.24E−01
HBr	0.00E+00	2.01E−03	8.06E−03
Br_2	0.00E+00	0.00E+00	0.00E+00
HMDS	0.00E+00	8.07E−02	3.23E−01
OMCTS	0.00E+00	0.00E+00	0.00E+00
Criteria			
CO	2.18E−01	0.00E+00	0.00E+00
NO	5.01E−03	0.00E+00	0.00E+00
NO_2	3.16E−03	0.00E+00	0.00E+00
SO_2	0.00E+00	0.00E+00	0.00E+00
N_2O	2.01E−04	2.64E−03	1.05E−02
O_3	0.00E+00	0.00E+00	0.00E+00

Table C.14 DRAM manufacturing emissions per wafer, 250 nm node

250 nm	Emissions to air		
kg/wafer	Expected value	Low	High
VOC			
DMA	8.75E−06	0.00E+00	0.00E+00
MMA	2.01E−04	0.00E+00	0.00E+00
TDMAT	0.00E+00	2.07E−03	8.29E−03
TMS	0.00E+00	0.00E+00	0.00E+00
TEOS	0.00E+00	0.00E+00	0.00E+00
formaldehyde	9.50E−05	9.60E−03	3.84E−02
m-cresol	6.02E−06	6.08E−03	2.43E−02
p-cresol	5.97E−06	6.03E−03	2.41E−02
PGMEA	1.12E−02	7.76E−02	3.11E−01
n-methyl-2-pyrollidone	0.00E+00	4.23E−03	1.69E−02
PGME	8.18E−03	1.02E−01	4.07E−01
ethyl lactate	9.54E−03	9.74E−02	3.89E−01
TMAH	1.05E−04	1.05E−02	4.20E−02
TDEAH	0.00E+00	0.00E+00	0.00E+00
TDMAS	0.00E+00	0.00E+00	0.00E+00
DEA	0.00E+00	0.00E+00	0.00E+00
PDMAT	0.00E+00	0.00E+00	0.00E+00

(continued)

Table C.14 (continued)

250 nm	Emissions to air		
kg/wafer	Expected value	Low	High
Flammable			
SiH_4	4.32E−04	6.16E−03	2.47E−02
H_2	7.13E−03	8.59E−02	3.44E−01
DCS	0.00E+00	0.00E+00	0.00E+00
B_2H_6	0.00E+00	0.00E+00	0.00E+00
C_4F_8	1.28E−04	0.00E+00	0.00E+00
GeH_4	0.00E+00	0.00E+00	0.00E+00
C_2H_2			
Other chemicals			
Ar	5.34E+00	2.67E+00	1.07E+01
N_2	1.33E+00	7.43E−01	2.97E+00
He	3.07E−03	1.53E−03	6.14E−03
O_2	4.42E−02	9.05E−01	3.62E+00
IPA	2.18E−02	2.08E+00	8.31E+00
NH_3	3.07E−03	1.53E−03	6.14E−03
Ti	7.76E−08	4.31E−08	1.72E−07
Cu	1.07E−07	1.28E−04	5.13E−04
Sn	4.06E−04	2.26E−04	9.03E−04

Table C.15 DRAM manufacturing emissions per wafer, 180 nm node

180 nm	Emissions to air		
kg/wafer	Expected value	Low	High
GWG			
CF_4	5.20E−04	4.69E−03	1.87E−02
CHF_3	4.60E−05	4.80E−04	1.92E−03
C_2F_6	0.00E+00	0.00E+00	0.00E+00
CH_4	1.46E−04	1.18E−01	4.73E−01
CO_2	6.51E−01	0.00E+00	0.00E+00
NF_3	3.77E−04	8.74E−02	3.50E−01
C_4F_6	0.00E+00	0.00E+00	0.00E+00
N_2O	0.00E+00	0.00E+00	0.00E+00
Hazardous			
F_2	2.98E−03	0.00E+00	0.00E+00
HF (gas)	2.52E−03	0.00E+00	0.00E+00
SiF_4	4.03E−05	3.07E−03	1.23E−02
C_4F_8	1.23E−06	6.78E−04	2.71E−03
C_2F_4	1.45E−05	0.00E+00	0.00E+00
COF_2	4.12E−06	0.00E+00	0.00E+00
HCl (gas)	3.97E−07	0.00E+00	0.00E+00
Cl_2	1.37E−04	1.82E−03	7.26E−03
$SiCl_4$	2.73E−06	0.00E+00	0.00E+00
PH_3	0.00E+00	4.57E−04	1.83E−03

(continued)

Table C.15   (continued)

180 nm	Emissions to air		
kg/wafer	Expected value	Low	High
BTBAS	0.00E+00	0.00E+00	0.00E+00
AsH_3	0.00E+00	0.00E+00	0.00E+00
BCl_3	5.57E−05	1.94E−03	7.77E−03
Pb	1.20E−04	5.98E−04	2.39E−03
WF_6	8.20E−03	3.10E−02	1.24E−01
HBr	0.00E+00	2.01E−03	8.06E−03
Br_2	0.00E+00	0.00E+00	0.00E+00
HMDS	0.00E+00	8.07E−02	3.23E−01
OMCTS	0.00E+00	0.00E+00	0.00E+00
Criteria			
CO	2.18E−01	0.00E+00	0.00E+00
NO	5.01E−03	0.00E+00	0.00E+00
NO_2	3.16E−03	0.00E+00	0.00E+00
SO_2	0.00E+00	0.00E+00	0.00E+00
N_2O	2.01E−04	2.64E−03	1.05E−02
O_3	0.00E+00	0.00E+00	0.00E+00

Table C.16   DRAM manufacturing emissions per wafer, 180 nm node

180 nm	Emissions to air		
kg/wafer	Expected value	Low	High
VOC			
DMA	8.75E−06	0.00E+00	0.00E+00
MMA	2.01E−04	0.00E+00	0.00E+00
TDMAT	0.00E+00	2.07E−03	8.29E−03
TMS	0.00E+00	0.00E+00	0.00E+00
TEOS	0.00E+00	0.00E+00	0.00E+00
formaldehyde	9.50E−05	9.60E−03	3.84E−02
m-cresol	6.02E−06	6.08E−03	2.43E−02
p-cresol	5.97E−06	6.03E−03	2.41E−02
PGMEA	1.12E−02	7.76E−02	3.11E−01
n-methyl-2-pyrollidone	0.00E+00	4.23E−03	1.69E−02
PGME	8.18E−03	1.02E−01	4.07E−01
ethyl lactate	9.54E−03	9.74E−02	3.89E−01
TMAH	1.05E−04	1.05E−02	4.20E−02
TDEAH	0.00E+00	0.00E+00	0.00E+00
TDMAS	0.00E+00	0.00E+00	0.00E+00
DEA	0.00E+00	0.00E+00	0.00E+00
PDMAT	0.00E+00	0.00E+00	0.00E+00

(continued)

Table C.16 (continued)

180 nm	Emissions to air		
kg/wafer	Expected value	Low	High
Flammable			
SiH$_4$	4.32E−04	6.16E−03	2.47E−02
H$_2$	7.13E−03	8.59E−02	3.44E−01
DCS	0.00E+00	0.00E+00	0.00E+00
B$_2$H$_6$	0.00E+00	0.00E+00	0.00E+00
C$_4$F$_8$	1.28E−04	0.00E+00	0.00E+00
GeH$_4$	0.00E+00	0.00E+00	0.00E+00
C$_2$H$_2$			
Other chemicals			
Ar	5.34E+00	2.67E+00	1.07E+01
N$_2$	1.33E+00	7.43E−01	2.97E+00
He	3.07E−03	1.53E−03	6.14E−03
O$_2$	4.42E−02	9.05E−01	3.62E+00
IPA	2.18E−02	2.08E+00	8.31E+00
NH$_3$	3.07E−03	1.53E−03	6.14E−03
Ti	7.76E−08	4.31E−08	1.72E−07
Cu	1.07E−07	9.36E−05	3.75E−04

Table C.17 DRAM manufacturing emissions per wafer, 130 nm node

130 nm	Emissions to air		
kg/wafer	Expected value	Low	High
GWG			
CF$_4$	2.28E−04	1.03E−03	4.13E−03
CHF$_3$	3.43E−06	5.13E−04	2.05E−03
C$_2$F$_6$	4.82E−07	0.00E+00	0.00E+00
CH$_4$	2.53E−04	1.29E−01	5.18E−01
CO$_2$	7.12E−01	0.00E+00	0.00E+00
NF$_3$	4.97E−04	1.15E−01	4.61E−01
C$_4$F$_6$	5.66E−06	3.61E−05	1.45E−04
N$_2$O	0.00E+00	0.00E+00	0.00E+00
Hazardous			
F$_2$	3.69E−03	0.00E+00	0.00E+00
HF (gas)	1.02E−03	0.00E+00	0.00E+00
SiF$_4$	4.01E−05	4.05E−03	1.62E−02
C$_4$F$_8$	2.11E−06	8.95E−04	3.58E−03
C$_2$F$_4$	1.92E−05	0.00E+00	0.00E+00
COF$_2$	5.73E−06	0.00E+00	0.00E+00
HCl (gas)	0.00E+00	0.00E+00	0.00E+00
Cl$_2$	1.24E−05	5.71E−04	2.28E−03
SiCl$_4$	0.00E+00	0.00E+00	0.00E+00

(continued)

Table C.17 (continued)

130 nm	Emissions to air		
kg/wafer	Expected value	Low	High
PH$_3$	0.00E+00	6.02E−04	2.41E−03
BTBAS	0.00E+00	0.00E+00	0.00E+00
AsH$_3$	0.00E+00	0.00E+00	0.00E+00
BCl$_3$	0.00E+00	0.00E+00	0.00E+00
Pb	4.56E−03	2.53E−03	1.01E−02
WF$_6$	1.08E−02	4.09E−02	1.64E−01
HBr	0.00E+00	2.66E−03	1.06E−02
Br$_2$	0.00E+00	0.00E+00	0.00E+00
HMDS	0.00E+00	1.16E−01	4.63E−01
OMCTS	0.00E+00	0.00E+00	0.00E+00
Criteria			
CO	4.32E−03	0.00E+00	0.00E+00
NO	4.08E−03	0.00E+00	0.00E+00
NO$_2$	7.90E−04	0.00E+00	0.00E+00
SO$_2$	0.00E+00	0.00E+00	0.00E+00
N$_2$O	2.65E−04	3.48E−03	1.39E−02
O$_3$	0.00E+00	0.00E+00	0.00E+00

Table C.18 DRAM manufacturing emissions per wafer, 130 nm node

130 nm	Emissions to air		
kg/wafer	Expected value	Low	High
VOC			
DMA	1.15E−05	0.00E+00	0.00E+00
MMA	2.66E−04	0.00E+00	0.00E+00
TDMAT	0.00E+00	2.73E−03	1.09E−02
TMS	0.00E+00	0.00E+00	0.00E+00
TEOS	0.00E+00	0.00E+00	0.00E+00
formaldehyde	1.36E−04	1.38E−02	5.50E−02
m-cresol	8.64E−06	8.73E−03	3.49E−02
p-cresol	8.55E−06	8.64E−03	3.46E−02
PGMEA	1.60E−02	1.11E−01	4.45E−01
n-methyl-2-pyrollidone	0.00E+00	5.58E−03	2.23E−02
PGME	1.17E−02	1.46E−01	5.84E−01
ethyl lactate	1.37E−02	1.40E−01	5.58E−01
TMAH	1.51E−04	1.51E−02	6.02E−02
TDEAH	0.00E+00	0.00E+00	0.00E+00
TDMAS	0.00E+00	0.00E+00	0.00E+00
DEA	0.00E+00	0.00E+00	0.00E+00
PDMAT	0.00E+00	0.00E+00	0.00E+00

(continued)

Table C.18 (continued)

130 nm	Emissions to air		
kg/wafer	Expected value	Low	High
Flammable			
SiH_4	5.70E−04	8.13E−03	3.25E−02
H_2	9.41E−03	1.13E−01	4.53E−01
DCS	0.00E+00	0.00E+00	0.00E+00
B_2H_6	0.00E+00	0.00E+00	0.00E+00
C_4F_8	3.75E−04	1.88E−04	7.50E−04
GeH_4	0.00E+00	0.00E+00	0.00E+00
C_2H_2			
Other chemicals			
Ar	7.05E+00	3.53E+00	1.41E+01
N_2	1.75E+00	9.80E−01	3.92E+00
He	4.05E−03	2.02E−03	8.09E−03
O_2	5.83E−02	1.14E+00	4.56E+00
IPA	3.12E−02	3.13E+00	1.25E+01
NH_3	4.05E−03	2.02E−03	8.09E−03
Ti	1.02E−07	5.69E−08	2.28E−07
Cu	1.41E−07	3.98E−04	1.59E−03
Sn	1.26E−03	7.00E−04	2.80E−03

Table C.19 DRAM manufacturing emissions per wafer, 90 nm node

90 nm	Emissions to air		
kg/wafer	Expected value	Low	High
GWG			
CF_4	2.00E−04	3.19E−03	1.28E−02
CHF_3	1.02E−05	1.75E−03	7.01E−03
C_2F_6	0.00E+00	0.00E+00	0.00E+00
CH_4	2.27E−04	2.45E−01	9.78E−01
CO_2	1.35E+00	0.00E+00	0.00E+00
NF_3	4.32E−04	1.04E−01	4.18E−01
C_4F_6	0.00E+00	0.00E+00	0.00E+00
N_2O	0.00E+00	0.00E+00	0.00E+00
Hazardous			
F_2	4.97E−03	0.00E+00	0.00E+00
HF (gas)	8.02E−04	0.00E+00	0.00E+00
SiF_4	5.78E−05	2.90E−03	1.16E−02
C_4F_8	1.63E−06	8.95E−04	3.58E−03
C_2F_4	1.92E−05	0.00E+00	0.00E+00
COF_2	5.05E−05	0.00E+00	0.00E+00
HCl (gas)	0.00E+00	0.00E+00	0.00E+00
Cl_2	2.49E−05	1.14E−03	4.57E−03

(continued)

Table C.19 (continued)

90 nm	Emissions to air		
kg/wafer	Expected value	Low	High
SiCl$_4$	0.00E+00	0.00E+00	0.00E+00
PH$_3$	0.00E+00	9.04E−04	3.61E−03
BTBAS	0.00E+00	0.00E+00	0.00E+00
AsH$_3$	0.00E+00	0.00E+00	0.00E+00
BCl$_3$	0.00E+00	0.00E+00	0.00E+00
Pb	4.56E−03	2.53E−03	1.01E−02
WF$_6$	6.49E−03	2.45E−02	9.81E−02
HBr	0.00E+00	5.32E−03	2.13E−02
Br$_2$	0.00E+00	0.00E+00	0.00E+00
HMDS	0.00E+00	1.43E−01	5.74E−01
OMCTS	0.00E+00	0.00E+00	0.00E+00
Criteria			
CO	2.38E−02	0.00E+00	0.00E+00
NO	3.52E−03	0.00E+00	0.00E+00
NO$_2$	8.10E−04	0.00E+00	0.00E+00
SO$_2$	0.00E+00	0.00E+00	0.00E+00
N$_2$O	5.30E−04	6.96E−03	2.78E−02
O$_3$	0.00E+00	0.00E+00	0.00E+00

Table C.20 DRAM manufacturing emissions per wafer, 90 nm node

90 nm	Emissions to air		
kg/wafer	Expected value	Low	High
VOC			
DMA	1.28E−06	0.00E+00	0.00E+00
MMA	2.95E−05	0.00E+00	0.00E+00
TDMAT	0.00E+00	3.04E−04	1.21E−03
TMS	1.96E−03	1.10E−03	4.40E−03
TEOS	1.37E−03	6.98E−04	2.79E−03
formaldehyde	1.58E−04	1.67E−02	6.66E−02
m-cresol	1.00E−05	1.90E−02	7.61E−02
p-cresol	9.92E−06	1.00E−02	4.01E−02
PGMEA	1.86E−02	1.41E−01	5.63E−01
n-methyl-2-pyrollidone	0.00E+00	1.68E−02	6.70E−02
PGME	1.35E−02	1.71E−01	6.83E−01
ethyl lactate	1.59E−02	1.62E−01	6.48E−01
TMAH	1.75E−04	1.75E−02	6.99E−02
TDEAH	0.00E+00	2.84E−05	1.13E−04
TDMAS	0.00E+00	0.00E+00	0.00E+00
DEA	0.00E+00	0.00E+00	0.00E+00
PDMAT	0.00E+00	0.00E+00	0.00E+00

(continued)

Table C.20 (continued)

90 nm	Emissions to air		
kg/wafer	Expected value	Low	High
Flammable			
SiH_4	4.51E−04	1.05E−02	4.19E−02
H_2	5.65E−03	6.75E−02	2.70E−01
DCS	0.00E+00	0.00E+00	0.00E+00
B_2H_6	0.00E+00	0.00E+00	0.00E+00
C_4F_8	1.20E−05	0.00E+00	0.00E+00
GeH_4	0.00E+00	0.00E+00	0.00E+00
C_2H_2			
Other chemicals			
Ar	5.60E+00	2.81E+00	1.12E+01
N_2	2.69E−01	1.67E−01	6.68E−01
He	4.50E−04	2.25E−04	8.99E−04
O_2	7.39E−02	9.89E−01	3.95E+00
IPA	3.61E−02	0.00E+00	0.00E+00
NH_3	4.50E−04	2.25E−04	8.99E−04
Ti	1.02E−07	5.69E−08	2.28E−07
Cu	1.41E−07	3.98E−04	1.59E−03
Sn	1.26E−03	7.00E−04	2.80E−03

Table C.21 DRAM manufacturing emissions per wafer, 70 nm node

70 nm	Emissions to air		
kg/wafer	Expected value	Low	High
GWG			
CF_4	1.89E−03	9.44E−04	3.78E−03
CHF_3	1.15E−05	5.73E−06	2.29E−05
C_2F_6	1.18E−05	5.91E−06	2.36E−05
CH_4	4.92E−03	2.46E−03	9.83E−03
CO_2	1.61E+00	8.06E−01	3.22E+00
NF_3	4.06E−04	2.03E−04	8.13E−04
C_4F_6	4.27E−05	2.14E−05	8.55E−05
N_2O	2.50E−04	1.25E−04	5.00E−04
Hazardous			
F_2	1.08E−03	5.38E−04	2.15E−03
HF (gas)	3.48E−02	1.74E−02	6.96E−02
SiF_4	9.73E−05	4.86E−05	1.95E−04
C_4F_8	4.32E−06	2.16E−06	8.64E−06
C_2F_4	8.05E−06	4.02E−06	1.61E−05
COF_2	1.33E−05	6.64E−06	2.66E−05
HCl (gas)	0.00E+00	0.00E+00	0.00E+00
Cl_2	2.35E−05	1.17E−05	4.69E−05
$SiCl_4$	0.00E+00	0.00E+00	0.00E+00
PH_3	0.00E+00	0.00E+00	0.00E+00

(continued)

Table C.21 (continued)

70 nm	Emissions to air		
kg/wafer	Expected value	Low	High
BTBAS	4.78E−06	2.39E−06	9.57E−06
AsH$_3$	0.00E+00	0.00E+00	0.00E+00
BCl$_3$	0.00E+00	0.00E+00	0.00E+00
Pb	0.00E+00	0.00E+00	0.00E+00
WF$_6$	2.04E−03	1.02E−03	4.08E−03
HBr	0.00E+00	0.00E+00	0.00E+00
Br$_2$	0.00E+00	0.00E+00	0.00E+00
HMDS	0.00E+00	0.00E+00	0.00E+00
OMCTS	0.00E+00	0.00E+00	0.00E+00
Criteria			
CO	6.00E−01	3.00E−01	1.20E+00
NO	6.78E−03	3.39E−03	1.36E−02
NO$_2$	6.46E−03	3.23E−03	1.29E−02
SO$_2$	0.00E+00	0.00E+00	0.00E+00
N$_2$O	2.50E−04	1.25E−04	5.00E−04
O$_3$	0.00E+00	0.00E+00	0.00E+00

Table C.22 DRAM manufacturing emissions per wafer, 70 nm node

70 nm	Emissions to air		
kg/wafer	Expected value	Low	High
VOC			
DMA	1.21E−06	6.05E−07	2.42E−06
MMA	2.78E−05	1.39E−05	5.57E−05
TDMAT	0.00E+00	0.00E+00	0.00E+00
TMS	2.05E−06	1.02E−06	4.09E−06
TEOS	1.88E−07	9.38E−08	3.75E−07
formaldehyde	1.44E−04	7.20E−05	2.88E−04
m-cresol	9.13E−06	4.56E−06	1.83E−05
p-cresol	9.04E−06	4.52E−06	1.81E−05
PGMEA	1.69E−02	8.47E−03	3.39E−02
n-methyl-2-pyrollidone	0.00E+00	0.00E+00	0.00E+00
PGME	1.23E−02	6.14E−03	2.46E−02
ethyl lactate	1.45E−02	7.23E−03	2.89E−02
TMAH	1.59E−04	7.95E−05	3.18E−04
TDEAH	0.00E+00	0.00E+00	0.00E+00
TDMAS	0.00E+00	0.00E+00	0.00E+00
DEA	0.00E+00	0.00E+00	0.00E+00
PDMAT	0.00E+00	0.00E+00	0.00E+00

(continued)

Table C.22 (continued)

70 nm	Emissions to air		
kg/wafer	Expected value	Low	High
Flammable			
SiH_4	4.17E−04	2.09E−04	8.34E−04
H_2	3.24E−03	1.62E−03	6.49E−03
DCS	0.00E+00	0.00E+00	0.00E+00
B_2H_6	0.00E+00	0.00E+00	0.00E+00
C_4F_8	3.18E−04	1.59E−04	6.36E−04
GeH_4	0.00E+00	0.00E+00	0.00E+00
C_2H_2			
Other chemicals			
Ar	1.77E+00	8.85E−01	3.54E+00
N_2	1.25E+00	6.25E−01	2.50E+00
He	1.35E−01	6.75E−02	2.70E−01
O_2	4.51E−02	2.26E−02	9.03E−02
IPA	3.31E−02	1.65E−02	6.61E−02
NH_3	1.35E−01	6.75E−02	2.70E−01
Ti	0.00E+00	0.00E+00	0.00E+00
Cu	0.00E+00	0.00E+00	0.00E+00
Sn	0.00E+00	0.00E+00	0.00E+00

Table C.23 DRAM manufacturing emissions per wafer, 57 nm node

57 nm	Emissions to air		
kg/wafer	Expected value	Low	High
GWG			
CF_4	8.44E−04	4.22E−04	1.69E−03
CHF_3	9.87E−06	4.93E−06	1.97E−05
C_2F_6	0.00E+00	0.00E+00	0.00E+00
CH_4	2.95E−04	1.48E−04	5.91E−04
CO_2	1.78E+00	8.92E−01	3.57E+00
NF_3	4.95E−04	2.47E−04	9.89E−04
C_4F_6	0.00E+00	0.00E+00	0.00E+00
N_2O	0.00E+00	0.00E+00	0.00E+00
Hazardous			
F_2	5.23E−03	2.61E−03	1.05E−02
HF (gas)	1.52E−03	7.61E−04	3.04E−03
SiF_4	6.55E−05	3.28E−05	1.31E−04
C_4F_8	1.19E−06	5.97E−07	2.39E−06
C_2F_4	1.41E−05	7.04E−06	2.82E−05
COF_2	5.17E−05	2.59E−05	1.03E−04

(continued)

Table C.23 (continued)

57 nm	Emissions to air		
kg/wafer	Expected value	Low	High
HCl (gas)	0.00E+00	0.00E+00	0.00E+00
Cl_2	4.10E−05	2.05E−05	8.21E−05
$SiCl_4$	0.00E+00	0.00E+00	0.00E+00
PH_3	2.49E−07	1.24E−07	4.97E−07
BTBAS	0.00E+00	0.00E+00	0.00E+00
AsH_3	3.11E−08	1.55E−08	6.21E−08
BCl_3	0.00E+00	0.00E+00	0.00E+00
Pb	0.00E+00	0.00E+00	0.00E+00
WF_6	1.02E−02	5.10E−03	2.04E−02
HBr	0.00E+00	0.00E+00	0.00E+00
Br_2	0.00E+00	0.00E+00	0.00E+00
HMDS	0.00E+00	0.00E+00	0.00E+00
OMCTS	0.00E+00	0.00E+00	0.00E+00
Criteria			
CO	3.85E−01	1.92E−01	7.69E−01
NO	9.16E−03	4.58E−03	1.83E−02
NO_2	4.82E−03	2.41E−03	9.64E−03
SO_2	0.00E+00	0.00E+00	0.00E+00
N_2O	5.00E−04	2.50E−04	1.00E−03
O_3	0.00E+00	0.00E+00	0.00E+00

Table C.24 DRAM manufacturing emissions per wafer, 57 nm node

57 nm	Emissions to air		
kg/wafer	Expected value	Low	High
VOC			
DMA	2.42E−06	1.21E−06	4.84E−06
MMA	5.57E−05	2.78E−05	1.11E−04
TDMAT	0.00E+00	0.00E+00	0.00E+00
TMS	1.85E−03	9.25E−04	3.70E−03
TEOS	1.29E−03	6.45E−04	2.58E−03
formaldehyde	1.59E−04	7.97E−05	3.19E−04
m-cresol	1.01E−05	5.05E−06	2.02E−05
p-cresol	1.00E−05	5.00E−06	2.00E−05
PGMEA	1.87E−02	9.37E−03	3.75E−02
n-methyl-2-pyrollidone	0.00E+00	0.00E+00	0.00E+00
PGME	1.35E−02	6.77E−03	2.71E−02
ethyl lactate	1.60E−02	8.00E−03	3.20E−02
TMAH	1.76E−04	8.81E−05	3.52E−04
TDEAH	0.00E+00	0.00E+00	0.00E+00
TDMAS	0.00E+00	0.00E+00	0.00E+00
DEA	0.00E+00	0.00E+00	0.00E+00
PDMAT	0.00E+00	0.00E+00	0.00E+00

(continued)

Table C.24 (continued)

57 nm	Emissions to air		
kg/wafer	Expected value	Low	High
Flammable			
SiH_4	1.22E−03	6.11E−04	2.45E−03
H_2	9.83E−03	4.92E−03	1.97E−02
DCS	0.00E+00	0.00E+00	0.00E+00
B_2H_6	0.00E+00	0.00E+00	0.00E+00
C_4F_8	2.27E−04	1.14E−04	4.55E−04
GeH_4	0.00E+00	0.00E+00	0.00E+00
C_2H_2			
Other chemicals			
Ar	1.44E+00	7.22E−01	2.89E+00
N_2	4.55E−01	2.27E−01	9.09E−01
He	8.48E−04	4.24E−04	1.70E−03
O_2	8.47E−02	4.23E−02	1.69E−01
IPA	3.65E−02	1.83E−02	7.30E−02
NH_3	8.48E−04	4.24E−04	1.70E−03
Ti	9.66E−08	4.83E−08	1.93E−07
Cu	6.66E−05	3.33E−05	1.33E−04
Sn	5.54E−03	2.77E−03	1.11E−02

C.3 DRAM LCA: Tabulated Results with Uncertainty

Table C.25 DRAM life cycle impacts per wafer, 57 nm node

	Photochemical smog kg NO$_x$			Acidification mol H$^+$			Ecotoxicity kg 2,4-D		
	Expected value	Lower bound	Upper bound	Expected value	Lower bound	Upper bound	Expected value	Lower bound	Upper bound
Transportation	0.33	N/A	N/A	10.7	N/A	N/A	0.00	N/A	N/A
Electricity (Si, fab. and use)	0.64	0.63	0.64	37	36	37	3.1	3.10	3.15
Fab. direct emissions	0.086	0.043	0.17	34	17	69	4.9E-02	2.5E-02	9.6E-02
Infrastructure	0.73	0.37	1.10	38	19.2	57	4.7E-03	2.4E-03	7.1E-03
Total	1.79	1.37	2.25	120	83	174	3.18	3.13	3.26

Table C.26 DRAM life cycle impacts per wafer, 57 nm node

	Human health cancer kg C_6H_6			Human health criteria total DALYs			Health noncancer kg C_7H_7		
	Expected value	Lower bound	Upper bound	Expected value	Lower bound	Upper bound	Expected value	Lower bound	Upper bound
Transportation	0.00	N/A	N/A	0.00	N/A	N/A	0.07	N/A	N/A
Electricity (Si, fab. and use)	0.00E+00	0.00E+00	0.00E+00	7.99E−03	7.95E−03	8.04E−03	251	249	253
Fab. direct emissions	1.01E−02	8.88E−03	1.14E−02	4.00E−05	2.00E−05	8.00E−05	885	776	996
Infrastructure	6.95E−03	3.48E−03	1.04E−02	1.30E−02	6.52E−03	1.96E−02	295	147	442
Total	0.017	0.012	0.022	0.022	0.0151	0.0283	1,431	1,173	1,691

	Eutrophication, to air kg N			Eutrophication, to water kg N		
	Expected value	Lower bound	Upper bound	Expected value	Lower bound	Upper bound
Transportation	1.18E−02	N/A	N/A	N/A	N/A	N/A
Electricity (Si, fab. and use)	2.27E−02	2.25E−02	2.30E−02	N/A	N/A	N/A
Fab. direct emissions	7.90E−04	3.95E−04	1.58E−03	5.11	2.61	10.11
Infrastructure	0.025	0.012	0.037	N/A	N/A	N/A
Total	0.060	0.047	0.074	5.1	2.6	10.1

Table C.27 DRAM life cycle impacts per wafer, 70 nm node

	Photochemical smog kg NO_x			Acidification mol H^+			Ecotoxicity kg 2,4-D		
	Expected value	Lower bound	Upper bound	Expected value	Lower bound	Upper bound	Expected value	Lower bound	Upper bound
Transportation	0.23	N/A	N/A	7.5	N/A	N/A	0.00	N/A	N/A
Electricity (Si, fab. and use)	0.49	0.48	0.49	28	28	28	2.4	2.37	2.42
Fab. direct emissions	0.064	0.032	0.13	18	9.0	36	3.3E−02	1.7E−02	6.5E−02
Infrastructure	0.61	0.31	0.92	32	16.1	48	3.9E−03	2.0E−03	5.9E−03
Total	1.40	1.05	1.78	85	60	120	2.43	2.39	2.5

Table C.28 DRAM life cycle impacts per wafer, 70 nm node

	Human health cancer kg C_6H_6			Human health criteria total DALYs			Health noncancer kg C_7H_7		
	Expected value	Lower bound	Upper bound	Expected value	Lower bound	Upper bound	Expected value	Lower bound	Upper bound
Transportation	0.00	N/A	N/A	0.00	N/A	N/A	0.05	N/A	N/A
Electricity (Si, fab. and use)	0.00E+00	0.00E+00	0.00E+00	5.95E−03	2.77E−03	−3.40E−03	192	190	194
Fab. direct emissions	9.18E−03	8.07E−03	1.03E−02	2.24E−05	1.12E−05	4.47E−05	815	715	915
Infrastructure	5.83E−03	2.91E−03	8.74E−03	1.09E−02	5.47E−03	1.64E−02	247	124	371
Total	0.015	0.011	0.019	0.017	0.0087	0.0135	1,254	1,029	1,480

	Eutrophication, to air kg N			Eutrophication, to water kg N		
	Expected value	Lower bound	Upper bound	Expected value	Lower bound	Upper bound
Transportation	8.24E−03	N/A	N/A	N/A	N/A	N/A
Electricity (Si, fab. and use)	1.74E−02	1.71E−02	1.77E−02	4.68	2.38	9.25
Fab. direct emissions	4.60E−04	2.30E−04	9.21E−04	N/A	N/A	N/A
Infrastructure	0.021	0.010	0.031	N/A	N/A	N/A
Total	0.047	0.036	0.058	4.7	2.4	9.2

Table C.29 DRAM life cycle impacts per wafer, 90 nm node

	Photochemical smog kg NO_x			Acidification mol H^+			Ecotoxicity kg 2,4-D		
	Expected value	Lower bound	Upper bound	Expected value	Lower bound	Upper bound	Expected value	Lower bound	Upper bound
Transportation	0.27	N/A	N/A	8.7	N/A	N/A	0.00	N/A	N/A
Electricity (Si, fab. and use)	0.51	0.50	0.51	29	29	29	2.5	2.47	2.52
Fab. direct emissions	0.072	0.036	0.14	31	16	63	2.2E−02	1.2E−02	4.2E−02
Infrastructure	0.52	0.26	0.77	27	13.5	40	3.3E−03	1.7E−03	5.0E−03
Total	1.37	1.06	1.70	96	67	141	2.52	2.48	2.6

Table C.30 DRAM life cycle impacts per wafer, 90 nm node

	Human health cancer kg C_6H_6			Human health criteria total DALYs			Health noncancer kg C_7H_7		
	expected value	Lower bound	Upper bound	Expected value	Lower bound	Upper bound	Expected value	Lower bound	Upper bound
Transportation	0.00	N/A	N/A	0.00	N/A	N/A	0.06	N/A	N/A
Electricity (Si, fab. and use)	0.00E+00	0.00E+00	0.00E+00	6.41E−03	3.74E−03	−1.43E−03	200	198	202
Fab. direct emissions	4.01E−02	2.84E−02	6.07E−02	2.69E−05	1.34E−05	5.37E−05	2,103	1,560	2,952
Infrastructure	4.89E−03	2.44E−03	7.33E−03	9.17E−03	4.58E−03	1.38E−02	207	104	311
Total	0.045	0.031	0.068	0.016	0.0088	0.0129	2,510	1,862	3,465

	Eutrophication, to air kg N			Eutrophication, to water kg N		
	Expected value	Lower bound	Upper bound	Expected value	Lower bound	Upper bound
Transportation	9.57E−03	N/A	N/A	N/A	N/A	N/A
Electricity (Si, fab. and use)	1.81E−02	1.79E−02	1.84E−02	4.67	2.38	9.23
Fab. direct emissions	5.44E−04	2.72E−04	1.09E−03	N/A	N/A	N/A
Infrastructure	0.018	0.009	0.026	N/A	N/A	N/A
Total	0.046	0.036	0.055	4.7	2.4	9.2

Table C.31 DRAM life cycle impacts per wafer, 130 nm node

	Photochemical smog kg NO_x			Acidification mol H^+			Ecotoxicity kg 2,4-D		
	Expected value	Lower bound	Upper bound	Expected value	Lower bound	Upper bound	Expected value	Lower bound	Upper bound
Transportation	0.17	N/A	N/A	5.4	N/A	N/A	0.00	N/A	N/A
Electricity (Si, fab. and use)	0.36	0.35	0.36	20	20	21	1.8	1.73	1.78
Fab. direct emissions	0.072	0.036	0.14	38	19	75	1.4E−02	7.7E−03	2.6E−02
Infrastructure	0.40	0.20	0.59	21	10.3	31	2.5E−03	1.3E−03	3.8E−03
Total	0.99	0.75	1.27	84	55	132	1.77	1.74	1.8

Table C.32 DRAM life cycle impacts per wafer, 130 nm node

	Human health cancer kg C_6H_6			Human health criteria total DALYs			Health noncancer kg C_7H_7		
	Expected value	Lower bound	Upper bound	Expected value	Lower bound	Upper bound	Expected value	Lower bound	Upper bound
Transportation	0.00	N/A	N/A	0.00	N/A	N/A	0.04	N/A	N/A
Electricity (Si, fab. and use)	0.00E+00	0.00E+00	0.00E+00	4.36E-03	2.30E-03	-1.65E-03	141	139	142
Fab. direct emissions	2.12E-02	1.46E-02	3.31E-02	3.20E-05	1.60E-05	6.41E-05	1,347	1,014	1,863
Infrastructure	3.75E-03	1.88E-03	5.63E-03	7.04E-03	3.52E-03	1.06E-02	159	80	239
Total	0.025	0.017	0.039	0.012	0.0061	0.0093	1,647	1,232	2,244

	Eutrophication, to air kg N			Eutrophication, to water kg N		
	Expected value	Lower bound	Upper bound	Expected value	Lower bound	Upper bound
Transportation	6.03E-03	N/A	N/A	N/A	N/A	N/A
Electricity (Si, fab. and use)	1.27E-02	1.25E-02	1.30E-02	3.87	1.95	7.70
Fab. direct emissions	6.37E-04	3.19E-04	1.27E-03	N/A	N/A	N/A
Infrastructure	0.013	0.007	0.020	N/A	N/A	N/A
Total	0.033	0.026	0.040	3.9	1.9	7.7

Table C.33 DRAM life cycle impacts per wafer, 180 nm node

	Photochemical smog kg NO$_x$			Acidification mol H$^+$			Ecotoxicity kg 2,4-D		
	Expected value	Lower bound	Upper bound	Expected value	Lower bound	Upper bound	Expected value	Lower bound	Upper bound
Transportation	0.07	N/A	N/A	2.4	N/A	N/A	0.00	N/A	N/A
Electricity (Si, fab. and use)	0.83	0.83	0.83	48	47	48	4.1	4.06	4.09
Fab. direct emissions	0.051	0.025	0.10	27	14	54	6.1E−03	3.6E−03	1.1E−02
Infrastructure	0.30	0.15	0.45	16	7.8	24	2.0E−03	1.0E−03	3.0E−03
Total	1.25	1.08	1.46	93	71	128	4.08	4.06	4.1

Table C.34 DRAM life cycle impacts per wafer, 180 nm node

	Human health cancer kg C_6H_6			Human health criteria total DALYs			Health noncancer kg C_7H_7		
	Expected value	Lower bound	Upper bound	Expected value	Lower bound	Upper bound	Expected value	Lower bound	Upper bound
Transportation	0.00	N/A	N/A	0.00	N/A	N/A	0.02	N/A	N/A
Electricity (Si, fab. and use)	0.00E+00	0.00E+00	0.00E+00	7.74E−03	6.19E−03	3.19E−03	327	326	328
Fab. direct emissions	1.27E−02	9.57E−03	1.80E−02	2.34E−05	1.17E−05	4.67E−05	718	563	948
Infrastructure	2.97E−03	1.48E−03	4.45E−03	5.33E−03	2.66E−03	7.99E−03	126	63	189
Total	0.016	0.011	0.022	0.013	0.009	0.011	1,171	952	1,465

	Eutrophication, to air kg N			Eutrophication, to water kg N		
	Expected value	Lower bound	Upper bound	Expected value	Lower bound	Upper bound
Transportation	2.63E−03	N/A	N/A	N/A	N/A	N/A
Electricity (Si, fab. and use)	2.96E−02	2.95E−02	2.97E−02	N/A	N/A	N/A
Fab. direct emissions	4.67E−04	2.33E−04	9.34E−04	2.76	1.41	5.45
Infrastructure	0.010	0.005	0.015	N/A	N/A	N/A
Total	0.043	0.037	0.049	2.8	1.4	5.5

Table C.35 DRAM life cycle impacts per wafer, 250 nm node

	Photochemical smog kg NO_x			Acidification mol H^+			Ecotoxicity kg 2,4-D		
	Expected value	Lower bound	Upper bound	Expected value	Lower bound	Upper bound	Expected value	Lower bound	Upper bound
Transportation	0.10	N/A	N/A	3.0	N/A	N/A	0.00	N/A	N/A
Electricity (Si, fab. and use)	1.02	1.01	1.02	58	58	59	5.0	4.99	5.02
Fab. direct emissions	0.051	0.025	0.10	27	14	54	7.9E−03	4.5E−03	1.5E−02
Infrastructure	0.26	0.13	0.39	14	6.8	21	1.7E−03	8.4E−04	2.5E−03
Total	1.43	1.27	1.61	102	82	137	5.01	4.99	5.03

Table C.36 DRAM life cycle impacts per wafer, 250 nm node

	Human health cancer kg C_6H_6			Human health criteria total DALYs			Health noncancer kg C_7H_7		
	Expected value	Lower bound	Upper bound	Expected value	Lower bound	Upper bound	Expected value	Lower bound	Upper bound
Transportation	0.00	N/A	N/A	0.00	N/A	N/A	0.02	N/A	N/A
Electricity (Si, fab. and use)	0.00E+00	0.00E+00	0.00E+00	9.53E−03	8.16E−03	5.55E−03	401	400	402
Fab. direct emissions	1.54E−02	1.13E−02	2.23E−02	2.34E−05	1.17E−05	4.67E−05	832	638	1,126
Infrastructure	2.48E−03	1.24E−03	3.72E−03	4.66E−03	2.33E−03	6.99E−03	105	53	158
Total	0.018	0.013	0.026	0.014	0.011	0.013	1,338	1,090	1,687

	Eutrophication, to air kg N			Eutrophication, to water kg N		
	Expected value	Lower bound	Upper bound	Expected value	Lower bound	Upper bound
Transportation	3.37E−03	N/A	N/A	N/A	N/A	N/A
Electricity (Si, fab. and use)	3.63E−02	3.62E−02	3.65E−02	N/A	N/A	N/A
Fab. direct emissions	4.67E−04	2.33E−04	9.34E−04	2.76	1.41	5.45
Infrastructure	0.009	0.004	0.013	N/A	N/A	N/A
Total	0.049	0.044	0.054	2.8	1.4	5.5

References

1. *Kirk-Othmer Encyclopedia of Chemical Technology*. John Wiley & Sons, 2001.
2. *Statistics for Industry Groups and Industries, Annual Survey of Manufacturers*. U.S. Census Bureau: Washington, DC, 2005.
3. 2008 Industry Overview, vol. 3, Semiconductors. Technical report, Standard and Poor, 2008.
4. *Ullmann's Encyclopedia of Industrial Chemistry, 7th Edition*. John Wiley & Sons, 2009.
5. Alliance to Save Energy. *PC Energy Report 2007, United States*. Unpublished report, available online: http://www.climatesaverscomputing.org/, 2007.
6. A. Andrae, P. Moller, J. Anderson, and J. Liu. Uncertainty estimation by Monte Carlo simulation applied to life cycle inventory of cordless phones and microscale metallization processes. *IEEE Transactions on Electronics Packaging Manufacturing*, 27(4):233–245, 2004.
7. A. S. Andrae, D. R. Andersson, and J. Liu. Significance of intermediate production processes in life cycle assessment of electronic products assessed using a generic compact model. *J. Cleaner Prod.*, 13(13-14):1269–1279, 2005.
8. C. P. Ausschnitt and A. C. Thomas. Advanced DUV photolithography in a pilot line environment,. *IBM Journal of Research and Development: Optical lithography*, 41(1), 1997.
9. S. C. Bartos, N. Kshetry, and C. S. Burton. Modeling China's semiconductor industry fluorinated compound emissions and drafting a roadmap for climate protection. *International Journal of Greenhouse Gas Control*, 2(4):665–676, Oct. 2008.
10. M. J. V. Beattie, W. G. Bacon, and R. Raudsepp. *Hydrometallurgical copper process. United States Patent, 4632738*. 1986.
11. M. Betz, M. Schuckert, and C. Herrmann. Life cycle engineering as decision making support in the electronics industry. In *Electronics and the Environment, 1998. ISEE-1998. Proceedings of the 1998 IEEE International Symposium on*, pages 231–236, 1998.
12. M. Blazek, S. Rhodes, F. Kommonen, and E. Weidman. Tale of two cities: environmental life cycle assessment for telecommunications systems. In *Proceedings of the IEEE International Symposium on Electronics and the Environment*, pages 76 – 81. IEEE, 11-13 May 1999.
13. P. Blowers and M. Titus. Use of life-cycle inventory as a screening tool for environmental performance: Supercritical carbon dioxide in the semiconductor industry. *Environmental Progress*, 23(4):284–290, 2004.
14. D. Bol, S. Boyd, and D. Dornfeld. Life-cycle energy demand of computational logic: From high-performance 32nm CPU to ultra-low-power 130nm MCU. *Proceedings of the 2011 IEEE International Symposium on Sustainable Systems and Technology (ISSST)*, 2011.
15a. S. Boyd, A. Horvath, and D. Dornfeld. Life-cycle energy and global warming potential of computational logic. *Environmental Science and Technology*, 43(19):7303–7309, 2009.
15b. S. Boyd, A. Horvath, and D. Dornfeld. Life-cycle assessment of computational logic produced from 1995 through 2010. *Environmental Research Letters*, 5(1), 2010.

15c. S. Boyd, A. Horvath, and D. Dornfeld. Life-cycle Assessment of Flash Memory. *IEEE Transactions on Semiconductor Manufacturing*, 24(1), 2011.

16. C. Brown and G. Linden. Offshoring in the semiconductor industry: Historical perspectives. *Brookings Trade Forum*, pages 279–322, 2005.

17. C. S. Burton and R. Beizaie. EPAs PFC Emissions Vintage Model (PEVM) v. 2.14: Description and documentation. Technical report, 2001.

18. Carnegie Mellon University Green Design Institute. *Economic Input-Output Life Cycle Assessment (EIO-LCA), US 1997 Industry Benchmark model, Economic Input-Output Life Cycle Assessment (EIO-LCA), US 1997 Industry Benchmark model.* http://www.eiolca.net, accessed May 15, 2009.

19. F.-T. Chang, Y.-C. Lin, H. Bai, and B.-S. Pei. Adsorption and desorption characteristics of semiconductor volatile organic compounds on the thermal swing honeycomb zeolite concentrator. *Journal of the Air & Waste Management Association*, 53(11):1384–1390, 2003.

20. H. Chein and T. M. Chen. Emission characteristics of volatile organic compounds from semiconductor manufacturing. *Journal of the Air & Waste Management Association*, 53(8):1029–1036, Aug. 2003.

21. D. Chen, X. Bi, J. Zhao, L. Chen, J. Tan, B. Mai, G. Sheng, J. Fu, and M. Wong. Pollution characterization and diurnal variation of PBDEs in the atmosphere of an e-waste dismantling region. *Environmental Pollution*, 157(3):1051–1057, Mar. 2009.

22. S. J. Chen. Beijing said to OK first Intel fab in China. *Forbes*, 3, 2007. (Accessed April 15 2008 http://www.forbes.com/).

23. Y. Chen and R. Liu. Future trend of flash memories. In *Memory Technology, Design and Testing, 2007. MTDT 2007. IEEE International Workshop on*, pages 17–18, 2007.

24. D. G. Coronell, T. H. L. Hsiung, H. P. J. Withers, and A. J. Woytek. *Process for nitrogen trifluoride synthesis. United States Patent, 5637285.* 1997.

25. C. Cremer, W. Eichhammer, M. Friedewald, P. Georgieff, S. Rieth-Hoerst, B. Schlomann, P. Zoche, B. Aebischer, and A. Huser. Energy consumption of information and communication technology in Germany up to 2010. Technical report, 2003.

26. J. G. Day. *Recovery of platinum group metals, gold and silver from scrap. United States Patent, 4427442.* 1984.

27. L. Deng and E. Williams. Measures and trends in energy use of semiconductor manufacturing. In *International Symposium on Electronics and the Environment*, Los Alamitos, CA, USA, 2008. IEEE.

28. W. Deng, J. Zheng, X. Bi, J. Fu, and M. Wong. Distribution of PBDEs in air particles from an electronic waste recycling site compared with Guangzhou and Hong Kong, South China. *Environment International*, 33(8):1063–1069, Nov. 2007.

29. X. Di, Z. Nie, B. Yuan, and T. Zuo. Life cycle inventory for electricity generation in China. *The International Journal of Life Cycle Assessment*, 12(4):217–224, June 2007.

30. D. Dickinson, J. Mosovsky, and S. Houthuysen. Assessing integrated circuit manufacturing for environmental performance and sustainability: A full scale IC business application. In *Proceedings of the 2003 IEEE International Symposium on Electronics and the Environment, IEEE*, 2003.

31. EC. Addressing the challenge of energy efficiency through information and communication technologies. Communication from the European Commission (COM 241) Technical Report COM(2008) 241 final, 2008.

32. EC. Ict and energy efficiency: The case for manufacturing. Technical report, Commission of the European Communities, Information Society & Media Directorate-General, Future and Emerging Technologies Proactive unit, 2009.

33. T. Ekvall and A. Andrae. Attributional and consequential environmental assessment of the shift to Lead-Free solders (10 pp). *The International Journal of Life Cycle Assessment*, 11(5):344–353, 2006.

34. EPA. Desktop computer displays: A life cycle assessment. Technical report, U.S. Environmental Protection Agency, Office of Pollution Prevention and Toxics, Design for Environment (DfE) Program, 2001.

35. EPA. Solders in electronics: A life-cycle assessment. Technical report, U.S. Environmental Protection Agency, Office of Pollution Prevention and Toxics, Design for Environment (DfE) Program, 2005.

36. EPA. EPA report to congress on server and data center energy efficiency. Technical report, U.S. Environmental Protection Agency, 2007.

37. EPA. *The Emissions & Generation Resource Integrated Database for 2007 (eGrid2007), available online: www.epa.gov/cleanenergy/energy-resources/egrid/index.html.* U.S. Environmental Protection Agency, September 2008.

38. EPA. *The Emissions & Generation Resource Integrated Database for 2007 (eGrid2007) Technical Support Document.* U.S. Environmental Protection Agency, September 2008.

39. EPA. Uses and emissions of liquid PFC heat transfer fluids from the electronics sector EPA-430-r-06-901. Technical report, U.S. Environmental Protection Agency, 2008.

40. C. Facanha and A. Horvath. Evaluation of life cycle air emission factors of freight transportation. *Environmental Science and Technology*, 41(20):7138–7144, 2007.

41. R. Faron, D. Devilliers, and R. Romano. Fluorine, in Ullmann's Encyclopedia of Industrial Chemistry *John Wiley & Sons*, 2000.

42. B. Foster, D. Meyersdorf, J. Padillo, and R. Brenner. Simulation of test wafer consumption in a semiconductor facility. In *IEEE/SEMI Advanced Semiconductor Manufacturing Conference and Workshop*, pages 298–302. IEEE, 1998.

43. R. E. Franklin, A. W. Francis, and G. Tarancon. *Process for the production of silane. United States Patent, 4041136.* 1977.

44. V. Fthenakis and H. Kim. Greenhouse-gas emissions from solar electric and nuclear power: A life-cycle study. *Energy Policy*, 35:2549–2557, 2007.

45. J. Gallarda and F. Wegrzyn. *Process and plant for production of carbon monoxide and hydrogen. United States Patent, 6098424.* 2000.

46. R. Galli, A.H. Bachmann, and H. Binder. Reduction of production costs through E4TM efficiency engineering. *Semiconductor Fabtech*, 8:85–90, 1998.

47. M. Geibig, J.; Socolof. Lead-free solders: A life-cycle assessment. Technical report, EPA744-R-05-001. Washington, DC., 2005.

48. GeSi. *Smart 2020: Enabling the low carbon economy in the information age.* unpublished report from the Global e-Sustainability Initiative, Available online at http://www.gesi.org/, 2008.

49. GeSi. *Using ICTs to Tackle Climate Change.* Available online at http://www.gesi.org/, 2008.

50. G. N. Gottier. *Low pressure stripping process for production of crude helium. United States Patent, 5011521.* 1991.

51. J. Harland, T. Reichelt, and M. Yao. Environmental sustainability in the semiconductor industry. In *Electronics and the Environment, 2008. ISEE 2008. IEEE International Symposium on*, pages 1–6, 2008.

52. W. G. Harris. *Method and apparatus for making hydrochloric acid and sodium hydroxide. United States Patent, 4749453.* 1988.

53. C. Hendrickson, A. Horvath, S. Joshi, and L. Lave. Economic input-output models for environmental life-cycle assessment. *Environmental Science and Technology*, 32(7):184A–191A, 1998.

54. C. T. Hendrickson, L. B. Lave, and H. S. Matthews. *Environmental Life Cycle Assessment of Goods and Services: An Input-Output Approach.* Resources for the Future Press, 2006.

55. P. Higginbotham, R. Agrawal, and D. M. Herron. *Process for the production of oxygen and nitrogen. United States Patent, 6227005.* 2001.

55x. T. Higgs, M. Cullen, M. Yao, and S. Stewart. Developing an Overall CO_2 Footprint for Semiconductor Products. *Proceedings of the IEEE International Symposium on Sustainable Systems and Technology, IEEE,* 2009.

56. H. Hoag. The missing greenhouse gas. *Nature Reports Climate Change*, (doi:10.1038), 2008.

57. A. C. Homme. *Contact sulfuric acid process employing double conversion/ double absorption. United States Patent, 4088742.* 1978.

58. J. Hsu, H. Lo, C. Pan, Y. Chen, and T. Hsieh. Test wafer control system in 300 mm fab. In *Semiconductor Manufacturing Technology Workshop Proceedings*, pages 33–36. IEEE, 2004.

59. S. C. Hu and Y. K. Chuah. Power consumption of semiconductor fabs in Taiwan. *Energy*, 28(8):895 – 907, 2003.

60. Intel. *Energy-Efficient Performance On The Client: A Proposed Measurement Methodology Version 2.0, available online http://www.intel.com/technology/eep/EEP_whitepaper.pdf.* Intel Corporation, Santa Clara, CA, 2007.

61. Intel. *Intel Corporation 5878 Corporate GHG Emissions Response SC09.* the Carbon Disclosure Project, Available online at: www.cdproject.net, 2009.

62. IPCC. *Good Practice Guidance and Uncertainty Management in National Greenhouse Gas Inventories: Chapter 6, Quantifying Uncertainties in Practice.* Intergovernmental Panel on Climate Change National Greenhouse Gas Inventories Programme (IPCC-TSU NGGIP), Japan, 2001.

63. IPCC. *Climate Change 2007: The Physical Science Basis, Contribution of Working Group I to the Fourth Assessment Report of the Intergovernmental Panel on Climate Change.* Cambridge University Press, Cambridge, UK and New York, NY, USA, 2007.

64. S. W. Jones. Can the semiconductor industry afford the cost of new fabs? accessed online: www.icknowledge.com/economics/fab_costs.html. *IC Knowledge*, 2004.

65. J. Kim, M. Xu, R. Kahhat, B. Allenby, and E. Williams. Designing and assessing a sustainable networked delivery (SND) system: Hybrid Business-to-Consumer book delivery case study. *Environmental Science and Technology*, 43(1):181–187, 2009.

66. S. Kim and M. Overcash. Energy in chemical manufacturing processes: gate-to-gate information for life cycle assessment. *Journal of Chemical Technology & Biotechnology*, 78(8):995–1005, 2003.

67. L. E. Kincaid, J. R. Geibig, and M. B. Swanson. Printed wiring board cleaner technologies substitutes assessment: Making holes conductive, EPA744-r-98-004a. Technical report, U.S. Environmental Protection Agency, 1998.

68. C. W. King and M. E. Webber. The water intensity of the Plugged-In automotive economy. *Environmental Science & Technology*, 42(12):4305–4311, June 2008.

69. E. Kitou and A. Horvath. External air pollution costs of telework. *The International Journal of Life Cycle Assessment*, 13(2):155–165, Mar. 2008.

70. N. Krishnan. *Design for Environment (DfE) in semiconductor manufacturing, Ph.D. Thesis.* PhD thesis, University of California at Berkeley, 2003.

71. N. Krishnan, S. Boyd, A. Somani, S. Raoux, D. Clark, and D. Dornfeld. A hybrid life cycle inventory of nano-scale semiconductor manufacturing. *Environmental Science and Technology*, 42(8):3069–3075, Apr. 2008.

72. N. Krishnan, R. Smati, S. Raoux, and D. Dornfeld. Alternatives to reduce perfluorinated compound (PFC) emissions from semiconductor dielectric etch processes: meeting environmental commitments while minimizing costs. In *Electronics and the Environment, 2003. IEEE International Symposium on*, pages 19–24, 2003.

73. N. Krishnan, S. Thurwachter, T. Francis, and P. Sheng. The environmental value systems (EnV-S) analysis: Application to CMP effluent treatment options. In *Proceedings of the Electrochemical Society (ECS) Conference on Improving Environmental Performance of Wafer Manufacturing Processes.* ECS, 2000.

74. N. Krishnan, E. Williams, and S. Boyd. Case studies in energy use to realize ultra-high purities in semiconductor manufacturing. In *Proceedings of the IEEE International Symposium on Electronics and the Environment.* IEEE, 2008.

75. LBNL. Energy efficient cleanroom information site, 2000.

76. S.-N. Li, J.-N. Hsu, H.-Y. Shih, S.-J. Lin, and J.-L. Hong. FT-IR spectrometers measure scrubber abatement efficiencies. *Solid State Technology*, 2002.

77. S. Lipp, G. Pitts, and F. Cassidy. *Environmental consciousness/: a strategic competitiveness issue for the electronics and computer industry.* Microelectronics and Computer Technology Corporation, Austin, TX, 1993.

78. Micron. *DDR2 Power Calc 16.XLS, Available online: download.micron.com.* Micron Semiconductor Products, 2007.

79. C. F. Murphy, G. A. Kenig, D. T. Allen, J.-P. Laurent, and D. E. Dyer. Development of parametric material, energy and emission inventories for wafer fabrication in the semiconductor industry. *Environmental Science and Technology*, 37(23):5373–5382, 2003.

80. N. Nissen, I. Griese, A. Middendorf, J. Muller, H. Potter, and H. Reichl. An environmental comparison of packaging and interconnection technologies. In *Electronics and the Environment, 1998. ISEE-1998. Proceedings of the 1998 IEEE International Symposium on*, pages 106–111, 1998.

81. G. A. Norris. Impact characterization in the tool for the reduction and assessment of chemical and other environmental impacts. *J. Ind. Ecol.*, 6(3-4):79–99, 2003.

82. J. Norton. Algebraic sensitivity analysis of environmental models. *Environ. Model. & Softw.*, 23(8):963–972, Aug. 2008.

83. M. O'Halloran. Fab Utility Cost Values for Cost of Ownership (CoO) Calculations, Technology Transfer #02034260A-TR, available online: www.sematech.org/docubase/abstracts/4260atr.htm. Technical Report 02034260A-TR, International Sematech, 2002.

84. S. Pacca and A. Horvath. Greenhouse gas emissions from building and operating electric power plants in the Upper Colorado River Basin. *Environmental Science and Technology*, 36(14):3194–3200, June 2002.

85. R. F. Pahade, J. B. Saunders, and J. J. Maloney. *Process for separating methane and nitrogen. United States Patent, 4592767.* 1986.

86. T. W. Peterson. VOC emissions reduction from photolithographic processes, available online: http://www.erc.arizona.edu/. Technical report, University of Arizona, 1999.

87. PG&E. Power content label, available online: www.pge.com, last accessed December 2008.

88. G. J. M. Phylipsen and E. A. Alsema. Environmental life-cycle assessment of multicrystalline silicon solar cell modules (report no. 95057). Technical report, Department of Science, Technology and Society, Utrecht University, the Netherlands, 1995.

89. S. M. Pieper, J. M. Paul, and M. J. Schulte. A new era of performance evaluation. *Computer*, 40(9):23–30, 2007.

90. E. Pinheiro, W.-D. Weber, and L. A. Barroso. Failure trends in a large disk drive population. In *Proceedings of the 5th USENIX Conference on File and Storage Technologies*, 2007.

91. A. Plepys. The grey side of ICT. *Environmental Impact Assessment Review*, 22:509–523, 2002.

92. A. Plepys. Software renting - better business, better environment: The case of application service providing (asp). In *Proceedings of the 2002 IEEE International Symposium on Electronics and the Environment*, pages 159–165. IEEE, 2002.

93. A. Plepys. The environmental impacts of electronics: Going beyond the walls of semiconductor fabs. In *Proceedings of the IEEE International Symposium on Electronics and the Environment*, pages 159–165. IEEE, 2004.

94. A. Plepys. *Environmental Implications of Product Servicing: The Case of Outsourced Computing Utilities.* PhD Thesis, Lund University, Sept. 2004.

95. A. Plepys. *Environmental Implications of Product Servicing: The Case of Outsourced Computing Utilities.* PhD thesis, Lund University, Sept. 2004.

96. M. J. Prather and J. Hsu. NF_3, the greenhouse gas missing from Kyoto. *Geophysical Research Letters*, 35:L12810, 2008.

97. S. Raoux, T. Tanaka, M. Bhan, H. Ponnekanti, M. Seamons, T. Deacon, L.-Q. Xia, F. Pham, D. Silvetti, D. Cheung, K. Fairbairn, A. Jonhson, R. Pearce, and J. Langan. Remote microwave plasma source for cleaning chemical vapor deposition chambers: Technology for reducing global warming gas emissions. *Journal of Vacuum Science & Technology B: Microelectronics and Nanometer Structures*, 17(2):477–485, 1999.

98. R. S. Reed. *Production of high purity hydrogen fluoride from silicon tetrafluoride. United States Patent, 4036938.* 1977.

99. C. Reich-Weiser, T. Fletcher, D. Dornfeld, and S. Horne. Development of the supply chain optimization and planning for the environment (SCOPE) tool - applied to solar energy. In *International Symposium on Electronics and the Environment.* IEEE, 2008.

100. J. Rosenblum, A. Horvath, and C. Hendrickson. Environmental implications of service industries. *Environmental Science and Technology*, 34(22):4669–4676, 2000.
101. K. W. Roth, F. Goldstein, and J. Kleinman. Energy consumption by office and telecommunications equipment in commercial buildings. volume i: Energy consumption baseline, adl 72895-00. Technical report, 2002.
102. K. Schischke, M. Stutz, J. Ruelle, H. Griese, and H. Reichl. Life cycle inventory analysis and identification of environmentally significant aspects in semiconductor manufacturing. In *Proceedings of the IEEE International Symposium on Electronics and the Environment*, pages 145–150. IEEE, 2001.
103. Sematech. The international technology roadmap for semiconductors, 2003 edition. Technical report, 1999.
104. Sematech. The international technology roadmap for semiconductors: 2009, 2007, 2005, 2003, 2001, 1999 editions and 2010, 2008 and 2006 updates. Technical report, 1999–2010.
105. Sematech. The international technology roadmap for semiconductors, 2003 edition. Technical report, 2001.
106. Sematech. The international technology roadmap for semiconductors, 2003 edition. Technical report, 2003.
107. Sematech. The international technology roadmap for semiconductors, 2005 edition. Technical report, 2005.
108. Sematech. The international technology roadmap for semiconductors, 2007 edition. Technical report, 2007.
109. SEMATECH. Guideline for environmental characterization of semiconductor process equipment, technology transfer document no. 06124825b-eng. Technical report, 2009.
110. J. M. Sherer. *Semiconductor industry wafer fab exhaust management*. CRC Press, 2005.
111. J. M. Sherer and J. W. Swauger. High removal efficiency water scrubbers for low inlet acid compound concentrations in the semiconductor industry, October 1998.
112. SIA. *Annual Forecast of Global Semiconductor Sales*. Semiconductor Industry Association, November 14 2007.
113. SIA. *Annual Forecast of Global Semiconductor Sales*. Semiconductor Industry Association, November 14 2008.
114. SICAS. *Semiconductor Wafer Fab Capacity and Utilisation Statistics Report, 3rd Quarter 2007,*. Semiconductor International Capacity Statistics (SICAS), Vessem, The Netherlands, 2007.
115. SICAS. *Semiconductor Wafer Fab Capacity and Utilisation Statistics Report, 3rd Quarter 2008,*. Semiconductor International Capacity Statistics (SICAS), Vessem, The Netherlands, 2008.
116. R. Smati, S. Raoux, and M. Woolston. Measuring and modeling gas consumption and emissions from semiconductor manufacturing processes. In *EHS Assessment Techniques, SEMICON West*, 2002.
117. S. Stewart. Reducing volatile organic compound emissions through improved waste management. In *Proceedings of the ISMI Symposium on Manufacturing Effectiveness*, 2005.
118. J. Stokes and A. Horvath. Life cycle energy assessment of alternative water supply systems. *The International Journal of Life Cycle Assessment*, 11(5):335–343, 2006.
119. F. Taiariol, P. Fea, C. Papuzza, R. Casalino, E. Galbiati, and S. Zappa. Life cycle assessment of an integrated circuit product. In *Proceedings of the IEEE International Symposium on Electronics and the Environment*, pages 128–133, 2001.
120. M. Tekwawa, S. Miyamoto, and A. Inaba. Life cycle assessment; an approach to environmentally friendly PCs. In *Proceedings of the IEEE International Symposium on Electronics and the Environment*, pages 125–130, 1997.
121. S. Thurwachter. *Environmental Value Analysis: Evaluating Manufacturing Product and Process Design Trade-Offs*. PhD thesis, University of California, Berkeley, 2000.
122. S. Thurwachter, J. Schoening, and P. Sheng. Environmental value (EnV) analysis. In *Electronics and the Environment, 1999. ISEE -1999. Proceedings of the 1999 IEEE International Symposium on*, pages 70–75, 1999.

123. M. W. Toffel and A. Horvath. Environmental implications of wireless technologies: News delivery and business meetings. *Environmental Science and Technology*, 38(11):2961–2970, June 2004.

124. P. Tuma and L. Tousignant. Reducing emissions of PFC heat transfer fluids. In *Proceedings of the 2001 SEMICON West, EHS Challenges and Analytical Methodologies session at the SEMI Technical Symposium: Innovations in Semiconductor Manufacturing, July 16, 2001*.

125. T. Ueno, T. Shiino, and H. Onishi. Evaluation of electronic components in life cycle assessment. *J Mater Cycles Waste Manag*, 1(1):25–32, 1999.

126. A. Watanabe, T. Kobayashi, T. Egi, and T. Yoshida. Continuous and independent monitor wafer reduction in DRAM fab. In *IEEE International Symposium on Semiconductor Manufacturing Conference Proceedings*, pages 303–306. IEEE, 1999.

127. J. L. Webster. *Preparation of tetrafluoroethylene*. Number 5684218. United States Patent No. 5684218, 1997.

128. B. P. Weidema and M. S. Wesnaes. Data quality management for life cycle inventories – an example of using data quality indicators. *J. Cleaner Prod.*, 4(3-4):167–174, 1996.

129. E. Williams. Energy intensity of computer manufacturing: Hybrid assessment combining process and economic input-output methods. *Environmental Science and Technology*, 38(22):6166–6174, Nov. 2004.

130. E. D. Williams, R. U. Ayres, and M. Heller. The 1.7 kilogram microchip: Energy and material use in the production of semiconductor devices. *Environmental Science and Technology*, 36(24):5504–5510, 2002.

131. S. M. Williams. The Santa Clara Valley Water District Fiscal Year 2007-08 Five-Year Capital Improvement Program. Technical report, Santa Clara Valley Water District, 2007.

132. M. Wong, S. Wu, W. Deng, X. Yu, Q. Luo, A. Leung, C. Wong, W. Luksemburg, and A. Wong. Export of toxic chemicals - a review of the case of uncontrolled electronic-waste recycling. *Environmental Pollution*, 149(2):131–140, Sept. 2007.

133. B. Y. Wu, Y. C. Chan, A. Middendorf, X. Gu, and H. W. Zhong. Assessment of toxicity potential of metallic elements in discarded electronics: A case study of mobile phones in china. *Journal of Environmental Sciences (China) [J. Environ. Sci. (China)]. Vol. 20*, 20(11):1403–1408, 2008.

134. M. Yao, A. Wilson, T. McManus, and F. Shadman. Comparative analysis of the manufacturing and consumer use phases of two generations of semiconductor microprocessors. In *Proceedings of the IEEE International Symposium on Electronics and the Environment*, pages 97–103. IEEE, 2004.

135. J. Yap. *The Asia-Pacific IT and High Technology Sectors: A Company and Industry Analysis*. Mergent, Inc., 2007.

136. J. C. Yung-Cheng and F. Cheng. Application development of virtual metrology in semiconductor industry. In *Proceedings of the IEEE Industrial Electronics Society Conference*. IEEE, 2005.

137. X. Zhou. *Life cycle thinking and assessment tools on environmentally benign electronics*. PhD thesis, University of California at Davis, 2007.

Index

A
Abatement
    facility abatement systems, 3, 73
    PFC abatement (*see* Per-fluorinated
        compound (PFC) abatement)
    point-of-use abatement, 74–75
    uncertainty, 83

B
Bipolar junction transistor (BJT),
    29–30

C
Carbon tetrafluoride (CF_4), 17, 64
Chip size
    CMOS, 53, 54, 62, 72
    DRAM, 98
    wafer fabrication, 20–21
Clean dry air (CDA), 53, 73
Clean room, fab systems, 52, 53
CMOS logic impact. *See* Life cycle impacts
Complementary metal oxide semiconductor
    (CMOS) logic
    air emissions and impacts
        acidification, 78–79
        EPA criteria human health,
          78, 80
        eutrophication, 78, 80
        smog formation, 78–79
    chemicals
        LCA data sources, 42–44
        production method, 41–42
    data quality, 67
    data sources, 38

electricity
    GWP, 46, 48
    primary energy use, 46, 48
global warming potential
    per computational power, 61
    per die, 76–77
human health impacts
    carcinogenicity, 81–82
    developmental/neurological toxicity, 80
    ecotoxicity, 81–82
    eutrophication, 78, 80
    life-cycle emissions, 80
    smog formation, 78–79
    worst-case eutrophication, 82–83
infrastructure and equipment, 46, 47
inventory model, 70
    energy and emissions, transportation,
      41
    GHG emission, EIO-LCA method, 40
    global warming potential, 38, 39
    life-cycle energy requirements, 38
    use-phase energy consumption, 41
    use-phase GWP emission, 41
manufacturing emissions data
    32 nm node, 139–141
    45 nm node, 137–139
    65 nm node, 135–137
    90 nm node, 133–135
    130 nm node, 131–133
    180 nm node, 129–131
    250 nm node, 127–129
    350 nm node, 125–127
manufacturing inventory data
    32 nm node, 124–125
    45 nm node, 122–123
    65 nm node, 121–122

S.B. Boyd, *Life-Cycle Assessment of Semiconductors*, DOI 10.1007/978-1-4419-9988-7,   221
© Springer Science+Business Media, LLC 2012

Complementary metal oxide semiconductor
(CMOS) logic (*cont.*)
90 nm node, 119–120
130 nm node, 118–119
180 nm node, 117–118
250 nm node, 115–116
350 nm node, 114–115
material demands and emissions, 70
methods
abatement, 73–75
electricity generation emissions factors,
75–76
end-of-life impacts, 72
environmental impact metrics, 75
facility utility systems and facility
energy efficiency, 73
life-cycle stages, data source types, 71
process, device and fab spreadsheet
models, 71–72
process power and emissions
manufacture, 72
use phase power, 76
per-wafer and per-die life-cycle impacts,
57–62
semiconductor manufacturing
fab model, mass and material
flows, 49
facility and process equipment energy
demand, 49, 52–54
per-fluorinated compound, 48
process emissions, 54
process flow, 49–51
sensitivity analysis
absolute sensitivity, 64–65
energy consumption, 65–67
GWP impacts, 67
relative sensitivity, 64–65
silicon, 44–45
transportation, 55, 56
uncertainty
abatement efficiency, 83
chemical production, 63–64
energy usage, 62, 63
GWP of emissions, 63, 64
sensitivity analysis, 83
unaccounted impacts, 84–85
use phase, 55–57
water, 45, 77–78

D
Data quality
CMOS, 67
life-cycle inventory model, 6

Die size
CMOS
energy usage, 58, 59, 62, 65
GWP, 59, 63, 64, 67
wafer fabrication, LCI, 20
Dynamic random access memory (DRAM),
31, 33, 35
environmental impacts
acidification, 101
associations, 99
ecotoxicity, 102
eutrophication via air emissions, 102
GWP, 100
infrastructure production, 102
smog formation, 101
human health impacts
carcinogenicity, 103
developmental, reproductive and
neurological toxicity, 103–104
life-cycle criteria, 103
manufacturing emissions data
57 nm node, 198–200
70 nm node, 196–198
90 nm node, 194–196
130 nm node, 192–194
180 nm node, 190–192
250 nm node, 188–190
manufacturing inventory data
57 nm node, 187–188
70 nm node, 185–186
90 nm node, 184–185
130 nm node, 182–183
180 nm node, 181–182
250 nm node, 179–180
manufacturing life cycle impacts
57 nm node, 201–202
70 nm node, 203–204
90 nm node, 205–206
130 nm node, 207–208
180 nm node, 209–210
250 nm node, 211–212
methodology, 98
resource consumption, 99
vs. Williams study
energy usage, 104–105
mass of, chemicals and gases, 106–107

E
Economic input-output life cycle assessment
(EIO-LCA) method
CMOS
chemicals, 42, 43
energy and greenhouse gas emission, 40

infrastructure and equipment, 46
uncertainty assessment, 63
wafer fabrication, 21
Electrically-erasable read-only memory
(EEPROM), 31
Electricity mix, 24, 46, 48
Energy consumption modeling
end-of-life, 26
high-performance logic, 24
life-cycle stage per die, 113
software, 26
system-level factors, 25
use phase scenario
electricity mix, 24
frequency and duration, 25
maximum rated power, 25
stand-by/idle, 25
utilization rate, 25
Environmental consciousness, 4–5
EPA
CMOS, 46
human health impacts, 80, 103
mass-balance method, 19
TRACI, 75
Equipment-centric inventory method, 6

F
Flash life cycle impacts
45 nm node, 169–170
65 nm node, 171–172
90 nm node, 173–174
120 nm node, 175–176
150 nm node, 177–178
Flash memory
environmental impacts, 90
GWP impacts, 92
vs. hard disk drives
life-cycle, 92–93
sensitivity analysis, 94–95
uncertainty, 94
human health impacts, 91
manufacturing emissions data
45 nm node, 167–168
65 nm node, 165–166
90 nm node, 163–164
120 nm node, 161–162
150 nm node, 159–160
manufacturing inventory data
45 nm node, 157–158
65 nm node, 156–157
90 nm node, 154–155
120 nm node, 153–154
150 nm node, 151–152

methodology, 88–89
NAND, 87, 88
primary energy consumption per memory
capacity, 89
process flow *vs.* single-level cells, 89
water consumption, 92, 93
Fourier transform infrared (FT-IR)
spectroscopy, 15–16

G
Ganzheitliche Bilanzierung (GaBi) LCA
software, 5
Global warming emissions. *See*
Complementary metal oxide
semiconductor (CMOS) logic
Global warming potential (GWP), 76–77
computational power, 61
electricity, 46, 48
energy-related emissions and direct GHG
release, 40
GHG emissions, 59
infrastructure and equipment, 46
life-cycle stage per die, 114
PFC emission, 59
process flows, 49
sensitivity analysis, 64, 66–67
transportation, 55, 58
uncertainty assessment, 62–64
use-phase electricity and energy
consumption, 41
Grid mix. *See* Electricity mix
Ground-level ozone formation. *See* Smog
formation
GWP. *See* Global warming potential
(GWP)

H
Hard disk drives *vs.* solid state drives
life-cycle, 92–93
sensitivity analysis, 94–95
uncertainty, 94
Heating ventilation and air conditioning
(HVAC), 53, 73
High purity chemicals, CMOS, 41, 42, 64
Human health impacts
CMOS logic
carcinogenicity, 81–82
developmental/neurological toxicity,
80
ecotoxicity, 81–82
life-cycle emissions, 80
worst-case eutrophication, 82–83

Human health impacts (*cont.*)
  DRAM
    carcinogenicity, 103
    developmental, reproductive and
      neurological toxicity, 103–104
    life-cycle criteria, 103

I
Information and communication technology
    (ICT), 1, 9
Infrastructure
  CMOS
    and equipment, total cost, 46, 47
    facility infrastructure, 40
    quality of data, 67
    Santa Clara Valley Water District
      infrastructure, 45
  life cycle inventory, 23
International Technology Roadmap for
    Semiconductors (ITRS), 52, 56, 60

K
Kyoto protocol, 54, 59

L
Life cycle impacts
  per die
    32 nm node, 142
    45 nm node, 143
    65 nm node, 144
    90 nm node, 145
    130 nm node, 146
    180 nm node, 147
    250 nm node, 148
    350 nm node, 149
  per wafer
    45 nm node, 169–170
    65 nm node, 171–172
    90 nm node, 173–174
    120 nm node, 175–176
    150 nm node, 177–178
Life cycle inventory (LCI)
  chemical production, 21–23
  energy consumption modeling
    end-of-life, 26
    software, 26
    system-level factors, 25
    use phase scenario, 24–25
  equipment-level, 14
  facility-level, 7, 10
  infrastructure, 23

  process-level, 7, 10
  wafer fabrication
    adaptation, 20–21
    EIO-LCA modeling, 21
    government and policy methods, 17–19
    mass and energy flow modeling, 14–17

M
Mass and material flow modeling, CMOS, 49
Million instructions per second (MIPS),
    56, 57

N
NF_3, PFC abatement, 17, 54, 59, 60

P
Per-fluorinated compound (PFC) abatement,
    48
  in China, 48
  GWG emission, 54
  Kyoto protocol, 54
  mass measurements, 54
  NF_3, 60
Photochemical oxidant formation (POF),
    78–80
Power consumption, CMOS, 55, 66, 67
Primary energy consumption
  CMOS, 46, 48
  DRAM, 97–99
  flash memory, 87–90
Process cooling water (PCW), 52, 73
Process flow, CMOS, 49–51

Q
Quadrupole mass spectrometry (QMS), 15

R
Restriction on Hazardous Substances (RoHS),
    2, 26, 72

S
Semiconductor industry
  CMOS, 27
  compound annual growth rate, 28
  dominant circuit elements, 29–30
  geographic concentration
    wafer manufacturing, 32–34
    Worldwide revenues, 33, 34

global annual revenues, 28, 29
per-fluoro-compounds, 27
product types
    analog electronics, 31
    analog ICs, 31
    discrete components, 31
    flash memory, 31
    micro-processing units, 31
    optical components, 31
    Worldwide revenues, 32
size, growth and trends, 28
technology node, 28–30
transportation, 27, 28
Semiconductor life-cycle assessment (LCA)
    challenges, 4
    CMOS logic (*see* Complementary metal
        oxide semiconductor
        (CMOS) logic)
    DRAM (*see* Dynamic random access
        memory)
    environmental impacts
        acid waste neutralization systems, 3
        embodied energy, 2
        end of life, 2
        fabrication facilities, 2
        facility abatement systems, 3
        use phase, 2
        wafer processing, 2–3
    flash memory (*see* Flash memory)
    future aspects
        IT *vs.* traditional products and services,
            111
        life-cycle inventory and impact factors,
            chemicals, 110–111
        networked thin client *vs.* local desktop
            computing, 111
        rebound effect, computational power
            and memory capacity, 112
    of information technology, 9–10
    process flow, 11
    timeline
        environmental consciousness, 4–5
        equipment-centric inventory method, 6
        externalization, 8
        GaBi LCA software, 5
        gate-to-gate LCI analysis, 6
        life-cycle inventory data, 5–6
        1.7 Kilogram Microchip, 7
Sensitivity analysis
    CMOS
        absolute sensitivity, 64–65
        energy consumption, 65–67
        GWP impacts, 67
        relative sensitivity, 64–65

flash memory, 94–95
human health cancer impacts, 83
Silicon, 44–45
Smog formation, 78–79
Solid state drives (SSD). *See* Flash memory
Substrate, 24, 29, 44
SYSMARK standard, 55

T
Taiwan Semiconductor Manufacturing
    Corporation (TSMC), 33
Technology node, 20
    CMOS
        building and equipment, total cost, 46,
            47
        chip sizes, 53, 54
        computational performance, 57
        facility equipment, 49, 52, 53
        GWP reduction, 49
        line yield, 53, 54
        production capacity fab, 37
        use phase power, 56
        wafer transportation and CO_2 emission,
            55, 56
        wafer yield, 53, 54
    semiconductor industry
        MOS production capacity, 30
        scaling and performance goals, 28–29
    wafer fabrication, LCI, 20
Tool for the Reduction and Assessment of
    Chemical and other environmental
    Impacts (TRACI), 75

U
Ultrapure water (UPW), 52, 73
United Nations Intergovernmental Panel on
    Climate Change (IPCC)
    global warming potentials, 75, 88, 98
    Tier 2a method, 18
    Tier 2c method, 19
Upstream environmental effects, CMOS logic
    chemicals, 41–44
    silicon, 44–45
    water, 45
Use phase scenario
    CMOS
        chip lifetime, 55
        CPU performance, 56–57
        energy consumption, 41
        GWP emissions, 41
        ITRS technology node, 56
        transistor size, 56

Use phase scenario (*cont.*)
  electricity mix, 24
  energy consumption modeling (*see* Energy
    consumption modeling)
  high-performance logic, 24
  power consumption, 76

V
Volatile organic compound (VOC),
  74

W
Wafer fabrication, LCI
  adaptation
    chip size, 20–21
    die size, 20
    product type, 20

    technology node, 20
    transistor configuration, 20
  EIO-LCA modeling, 21
  government and policy methods, 17–19
  integrated circuit, 13, 14
  mass and energy flow modeling
    electricity generation, emissions factors,
      17
    facility utilities and equipment design,
      16–17
    semiconductor process emissions
      measurement, 15–16
Water consumption, 77–78
  electricity generation, 92, 94, 99
  life-cycle stage
    per die, 113
    per memory capacity, 93, 100
Water supply, CMOS, 45
World Semiconductor Council (WSC), 52